HOW TO PASS
THE CALIFORNIA
ENTRY LEVEL
MATHEMATICS
TEST

HOW TO PASS THE CALIFORNIA ENTRY LEVEL MATHEMATICS TEST

VIGGO P. HANSEN
California State University, Northridge

EVELYN D. BELL
Ysleta Independent School District

ROBERT J. WISNER
New Mexico State University

JOHN WILEY & SONS
New York   Chichester   Brisbane   Toronto   Singapore

Library of Congress Cataloging in Publication Data:

Hansen, Viggo P.
    How to pass the California entry level mathematics test.

    Bibliography: p.
    Includes index.
    1. Mathematics--Examinations, questions, etc.
2. Mathematics--Problems, exercises, etc.   I. Bell,
Evelyn.   II. Wisner, Robert J.   III. Title.
IV. Title: California entry level mathematics test.
QA43.H264   1987   510'76   86-15926
ISBN 0-471-82779-7 (pbk.)

Printed in the United States of America

10 9 8 7 6 5 4 3 2 1

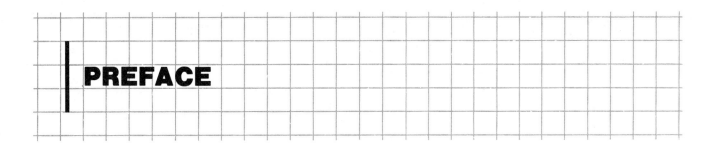

PREFACE

This book was written for the large number of college or college-bound students who need to review or to learn for the first time the concepts and skills of basic mathematics—arithmetic, algebra, and geometry—that are essential for success in the California Entry Level Mathematics (ELM) Test.

The contents of the book adhere faithfully to the outline prepared by The California State University in its *Entry Level Mathematics Workbook* (January 1984).

This Preface contains the need for taking tests and immediately unveils the book's goal: passing the ELM test.

The book then proceeds to explore the topics that were presented in the official California *Entry Level Mathematics Workbook*. Our presentation follows the workbook's tacit prescription: one idea and one skill at a time.

FORMAT

The book's mode of presentation is highly structured. The ELM main topics (Arithmetic Skills, Elementary Algebra Skills, and Geometric Measurement Skills) are what we call *Topics*. Each Topic is broken into Chapters, and each Chapter is broken into Units. Each unit comprises either two or four pages of explanation with numerous teaching examples followed by a similar number of pages of problems. Almost all of the units are of the shorter variety to help maintain the user's interest and attention.

EXAMPLES

Most of the teaching and learning experiences rely on the examples. There is, in fact, minimal discourse apart from the examples. The example sets begin with very easy material and progress in level, sometimes ending with the solutions of word problems in mathematics. The examples of word problems appear as they are called for to fit the material at hand.

PROBLEMS

In each unit there is example–problem conformity: the problems parallel the examples, so that the first few problems are like Example A, the next few like Example B, and so forth. There are also occasionally—and as appropriate—some examples and problems that lead to a general mathematical principle that is useful for many levels of mathematics.

This example–problem conformity means that the problems are graded. The grading provides for many successful experiences for the student. And as with the examples, problem sets from time to time end with word problems.

The problems are "paired," where the odd-numbered problems provide essentially the same experiences as the even-numbered problems.

The pages with problems are perforated so that they may be submitted to an instructor. There is never a place where the removal of a problem page removes any teaching page so that the book will always be useful as a reference for basic mathematics.

ACHIEVEMENT TESTS

Throughout the book, there are achievement tests that are designed to identify weaknesses that the student may still possess despite the interposed instruction. An inability to do any one of the achievement test items signals a need for further study of the material involved.

The achievement tests conform to the multiple-choice structure of the ELM testing program. Indeed, they are designed to look like ELM tests so that the students can become comfortable with the appearance of ELM tests.

ANSWERS

Answers for odd- and even-numbered problems, as well as selected solutions to achievement tests, are given at the end of the text.

GLOSSARY

There is a glossary at the back of the book, just before the index. As one would expect, the glossary contains the technical terms used in the book, along with short definitions. But more than that, the glossary also contains some terms not used in the book. The purpose of including the additional terms is to serve those instructors who may choose to use them in their presentations.

INSTRUCTOR'S EDITION

In place of an "instructor's manual" that is commonly provided for adopters of a book, we have prepared a separate edition. This contains a page-by-page reduced reproduction of the student edition with all the answers on the page and numerous marginal notes. The marginal notes selectively include:

1. An *objective* for each unit.

2. *Suggestions* to the instructor concerning presentation of the unit material.

3. Space for *notes* that the instructor may wish to make.

4. *Additional examples* that can be used to expand the presentation of the unit.

5. *Extended examples* that require multiple-step processes.

6. *Application examples*.

7. Some *selected solutions* to problems that emphasize the skills needed in the unit.

8. *Extended problems* similar to the extended examples, with answers.

9. *Applications*, problems similar to the application examples, with answers.

10. *Extra problems*, with answers.

11. *Supplementary references*, listing materials that are available from John Wiley & Sons that relate to the unit material.

Also included in the instructor's edition are Mid-Book Tests and End-of-Book Tests, two versions of each: one that is multiple choice and one that is not.

OTHER SUPPLEMENTARY MATERIALS

It is clear that the instructor's edition is a major supplement, including features that would normally be given in a list of supplementary materials. In addition, there are the following:

A *computerized testgenerator* that will generate tests, quizzes, and exams, which is available for the Apple or IBM computer. This ELM software is designed exclusively to serve the needs of the users of this book.

A *printed testbank* in which there are numerous tests, both "open ended" and multiple choice.

USES

There are various ways that this book can be used. It is ideal for many modes of instruction, including its use as:

A *self-study manual* for those who wish to prepare for the ELM test on their own initiative.

A *textbook for a self-paced course.*

A *textbook in a classroom setting*, used in the manner that textbooks are ordinarily used.

A *mathematics learning center textbook* where the modes of instruction vary widely from school to school, but for which this book may be adapted to meet the needs of almost any mode of learning center instruction.

A *tutorial or directed study textbook*, wherein there are no formal classroom lectures.

COST

We are pleased with the publisher's attention to the many details involved in keeping the price of the book as low as possible; that John Wiley & Sons did this without sacrificing quality is greatly appreciated.

Northridge, California   Viggo P. Hansen
El Paso, Texas   Evelyn D. Bell
Las Cruces, New Mexico   Robert J. Wisner

TO THE STUDENT

Taking tests in mathematics has been a fact of life ever since the inception of formal education. This occurred for at least two reasons: (1) answers on mathematics tests are usually irrefutably right or wrong, and (2) there is an accepted body of information that constitutes mathematics in selected categories—like arithmetic, geometry, and algebra—so it is easy to construct a test.

Another reason for testing, which is less often appreciated by students, is that a test helps you to assess your own level of mathematical understanding; that is, it gives you an honest evaluation. This evaluation then provides the information as to where you need to study. It assists in making the decision as to where your time and effort can be most efficiently spent. For this reason, we created this textbook.

We began by accepting the major themes and objectives in arithmetic, algebra, and geometry that the California State University and College system has identified as essential for university admission. Having successfully demonstrated your ability to handle these concepts as measured by the ELM test, you should be able to handle the minimum mathematical skills needed for the bachelor's degree. We considered these objectives as the foundation for the mathematics content of this book.

Next, in order to save your study time and energy, we developed test items that parallel those included on the ELM test. If you are able to answer these, you will have little trouble when taking the ELM tests. However, should some of the items give you a problem, you are furnished with many examples that show you how to solve the problems. These examples have been very carefully selected to follow the logical growth pattern in your understanding of the specific concepts. Even though you may be able to answer the test items correctly, you are still encouraged to study the examples to ensure that you have not missed some of the subtle features in each problem.

When you take a test that will have a powerful impact on your life, such as gaining university admission, you cannot be overly prepared. The pressure of a test puts stress on your ability to solve problems quickly and correctly. The better you understand the various mathematical processes being tested, the less affected you will be by the testing environment; in other words, you will be confident in your abilities. You are encouraged to take the included test items under timed conditions. That is the way the ELM test is given, and being prepared for this will greatly improve your score.

Keep in mind that your future success in mathematics is dependent upon your present understanding. If you are able to complete this book and the tests satisfactorily, your university experience with mathematics will be more rewarding and your selected career more successful. When you have passed the ELM test, you will feel confident that you are capable of taking higher level courses in mathematics. We encourage you to study the first section for specific suggestions on how to take the ELM mathematics test.

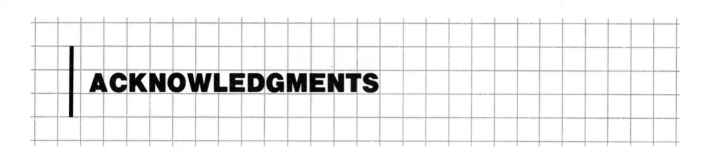

ACKNOWLEDGMENTS

The authors could not have produced this book without a great deal of assistance. First and foremost, for her help and encouragement above and beyond the call of duty, we thank Carolyn Moore, mathematics editor of John Wiley & Sons. In addition, we extend our thanks to Karin Gerdes Kincheloe, senior designer, Barbara Heaney, senior editor, Glenn Petry, copy editor, and Elizabeth Meder, senior production supervisor. We would also like to thank the following reviewers: Barbara Buhr, Fresno City College; William E. Colburn, Jr., Mira Costa College; Meriam Keesey, San Diego State University, and Edsel Stiel, California State University–Fullerton.

V.P.H.
E.D.B.
R.J.W.

HINTS ON TAKING THE ELM TEST

Under the rules as of the publication date of this book, students who need to pass the ELM test must take it during their first or, for schools on the quarter system, during their first or second quarter in the State University System. You will need to check on the dates and places that the test will be given at your school.

The examination will test your skills in arithmetic, elementary algebra, and geometric measurement. That is why this book is presented in terms of Topics A, B, and C, covering precisely those respective skills.

The ELM test items are multiple choice problems, with five choices for each item. You will be allotted 75 minutes. Approximately 20% of the test items are from arithmetic, approximately 60% from algebra, and approximately 20% from geometry. The passing score is independent of those percentage allocations; that is to say, you could get most of your correct responses from arithmetic and algebra, for example. Or you could get a lot of those from arithmetic and geometry.

The purpose of these hints on taking the test is to help you maximize your score. Many students have found these hints (along with possibly some others) to be useful, and we hope you will too.

1. *Avoid test anxiety*

Try to approach the examination in a frame of mind that will help you to enhance your score. Try not to let other concerns get in the way of your mental acuity. For example, many people will advise you to get a good night's rest before the test. Others will advise you to cram for the test so as to wipe out other concerns. Visit a restroom before the test so you will not interrupt your test taking by an emergency call of nature. You must decide how much of this sort of advice best suits your personality and temperament. Think about it.

2. *Understand the instructions*

Be sure that you understand both written and any oral instructions, including what sort of pencils you might need to bring, how to record your answers, and so forth. If you have any questions about the instructions, get them answered before the test begins.

3. *Watch your time!*

You will have about one minute and ten seconds per test question. That may sound like a lot of time, and since you will undoubtedly get many of your answers in much less time, you will have plenty of time for the others. On the other hand, 75 minutes will pass more rapidly than you think. Thus, take a watch with you to keep track of your time.

4. *Do the easy problems first*

Do not necessarily do the problems according to the order in which they occur. Do first the ones in which you *know* the answer right away. *CAUTION*: In working the problems this way, be very sure that your response item number on the answer sheet corresponds to your test item number.

5. *Go through a second time*

After answering the items you are sure of, go through the test a second time, working those problems that you know *how* to do but which need your further attention in order to get the right answer.

6. *Go through a third time*

Now go through the test again. The problems that remain are those involving either confusion or uncertainty on your part. Omit the choice or choices for each item that are clearly incorrect. There will usually be at least one or two such cases that will occur to you for each remaining test item.

7. *Guess when needed*

For each item that arises in this third time through, simply *guess* among the three or four choices that remain. *You have nothing to lose by guessing*! Many tests have a scoring system that will penalize you for a wrong answer, but the ELM scoring system does not.

8. *Answer each question*

For the reasons just given, make *very* sure that you have an answer for each and every test item. In the worst case possible—that you guess one from among all five choices—you have a 20% chance of getting the right answer. If you can omit even one unreasonable choice, you have a 25% chance, and if you can omit two unreasonable choices, you have a one-in-three chance. It's worth it to guess on these test items.

9. *Use all of your time*

Be sure to use all 75 minutes of your time. You will notice that some students will leave early. Do not let this distract you. They are being unfair to themselves for not taking advantage of the next hint.

10. *Check all the answers*

Look at your remaining time. Spend *all* of that time in checking over your answers, beginning with making sure that you have an answer for each item. If you detect a wrong response on your part and need to change your answer because you are *sure* that you are wrong, be careful to erase fully that response on your answer sheet and replace it appropriately. But if you just guessed at an answer and only *think* that you *might* be wrong, be careful! A first guess might well have a better chance than a second guess.

Moses presented humankind with ten commandments. We have given you ten hints. Use them as you see fit.

One final thought: Good luck to you! We hope you pass the test, and we also hope that your study of this book will assure your success.

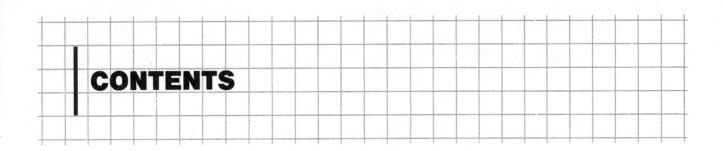

CONTENTS

TOPIC C   GEOMETRIC MEASUREMENT SKILLS

ARITHMETIC SKILLS

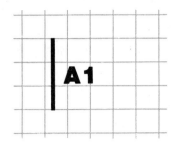

WHOLE NUMBERS AND THEIR OPERATIONS

A1.1 PLACE VALUE

Numbers are written using just ten digits. The digits we use for writing numbers are 0, 1, 2, 3, 4, 5, 6, 7, 8, and 9. To write numbers larger than 9, we use rules to govern the position, or **place value**, of each digit in the number. The position of a digit in a number determines its value in that number.

	Billions	Hundred millions	Ten millions	Millions	Hundred thousands	Ten thousands	Thousands	Hundreds	Tens	Ones
five hundred thirteen								5	1	3
two thousand three hundred twenty five							2	3	2	5
one thousand six							1	0	0	6
twenty seven thousand nine hundred twenty five						2	7	9	2	5
seven hundred sixteen thousand four hundred thirty three					7	1	6	4	3	3
thirty seven million nine thousand six hundred		3	7	0	0	9	6	0	0
one billion seven thousand	1	0	0	0	0	0	7	0	0	0
four hundred million four hundred forty four		4	0	0	0	0	0	4	4	4
nine hundred sixty two thousand one hundred sixty two					9	6	2	1	6	2
five billion	5	0	0	0	0	0	0	0	0	0

Note the use of zero (0) as a place holder when no value for a position was given.

Example A   749 means $\underline{7}$ hundred + $\underline{4}$ tens + $\underline{9}$ ones.

Example B   9603 means $\underline{9}$ thousands + $\underline{6}$ hundreds + $\underline{0}$ tens + $\underline{3}$ ones.

Example C $\underline{5}$ hundreds, $\underline{3}$ tens, $\underline{2}$ ones is the same as $\underline{532}$.

Example D $\underline{6}$ thousands, $\underline{8}$ ones, $\underline{5}$ tens, $\underline{0}$ hundreds is the same as $\underline{6058}$.

Example E $\underline{7}$ hundreds, $\underline{5}$ thousands, $\underline{6}$ tens, $\underline{3}$ ten thousands, $\underline{0}$ ones is the same as $\underline{35760}$.

To make it easier to read large numbers, commas are used to group the digits. To insert commas, begin at the right of the number and move to the left, inserting a comma in front of each set of three digits, but not in front of the entire number.

Example F   1749632 is written with commas as 1,749,632.

Example G   432875961 is written with commas as 432,875,961.

Since we write numbers using the ten digits, our numeration system is called a **base ten** system. We can see how this operates by writing our numbers in what is called **expanded form**.

Example H   43 = 40 + 3 or four tens plus three ones, or forty plus three.

Example I   7934 = 7000 + 900 + 30 + 4 or seven thousand plus nine hundred plus thirty plus four.

It is possible to continue this expansion of numbers using multiples of 10 for place values.

Example J $3478 = 3000 + 400 + 70 + 8$
$= (3 \times 1000) + (4 \times 100) + (7 \times 10) + (8 \times 1).$

Example K $40,650 = (4 \times 10,000) + (6 \times 100) + (5 \times 10).$

When using the expanded form, it is not necessary to write anything when there is a 0 in a given place. However, it is not incorrect to do so, as illustrated below with the number from Example K.

Example L $40,650 = (4 \times 10,000) + (0 \times 1000) + (6 \times 100) + (5 \times 10) + (0 \times 1).$

ERIC Bellamy   23R   09/26/88

NAME   SECTION   DATE   INSTRUCTOR

A1.1 PROBLEMS

Complete the following.

1. 482 = ___4___ hundreds + ___8___ tens + ___2___ ones

2. 583 = ___5___ hundreds + ___8___ tens + ___3___ ones

3. 940 = ___9___ hundreds + ___4___ tens + ___0___ ones

4. 437 = ___4___ hundreds + ___3___ tens + ___7___ ones

5. 300 = ___3___ hundreds + ___0___ tens + ___0___ ones

6. 7398 = ___7___ thousands + ___3___ hundreds + ___9___ tens + ___8___ ones

7. 9732 = ___9___ thousands + ___7___ hundreds + ___3___ tens + ___2___ ones

8. 1111 = ___1___ thousands + ___1___ hundreds + ___1___ tens + ___1___ ones

9. 2040 = ___2___ thousands + ___0___ hundreds + ___4___ tens + ___0___ ones

10. 8302 = ___8___ thousands + ___3___ hundreds + ___0___ tens + ___2___ ones

11. 9999 = ___9___ thousands + ___9___ hundreds + ___9___ tens + ___9___ ones

12. 3004 = ___3___ thousands + ___0___ hundreds + ___0___ tens + ___4___ ones

13. 5000 = ___5___ thousands + ___0___ hundreds + ___0___ tens + ___0___ ones

14. 1234 = ___1___ thousands + ___2___ hundreds + ___3___ tens + ___4___ ones

15. 2863 = ___2___ thousands + ___8___ hundreds + ___6___ tens + ___3___ ones

Rewrite the following using digits only.

16. 5 hundreds, 3 tens, 2 ones = __532__

17. 8 hundreds, 0 tens, 7 ones = __807__

18. 1 thousands, 3 tens, 0 ones = __1030__

19. 9 hundreds, 3 ones, 8 tens, 6 thousands = __6983__

20. 3 thousands, 6 ones, 5 hundreds, 0 tens = __3506__

Insert commas, then write the following numbers with words.

21. 8,462,900,473 _8 Billions, 4 hundred millions, 6 Ten millions, 2 millions, 9 Hundred Thousands, 4 hundreds, 7 Tens, 3 ONES_

22. 306,306,306 _3 Hundred millions, 6 Ten millions, 3 Hundred Thousands, 6 Thousands, 3 Hundreds, 6 ONES_

Write the following in expanded form with multiples of 10 for place values.

23. 58 _(5×10) + (8×1)_

24. 890 _(8×100) + (9×10)_

25. 1500 _(1×100) + (5×100)_

26. 3002 _(3×1000) + (2×1)_

27. 30,890 _(3×10,000) + (8×100) + (9×10)_

28. 99,300 _(9×10,000) + (9×1000) + (3×100)_

29. 105,900 _(1×100,000) + (5×1000) + (9×100)_

30. 300,000 _(3×100,000)_

31. 8,903,000 _(8×1,000,000) + (9×100,000) + (3×1000)_

32. 58,000,005 _(5×10,000,000) + (8×1,000,000) + (5×1)_

33. 7,009,600,300 _(7×1,000,000,000) + (9×1,000,000) + (6×100,000) + (3×100)_

A1.2 ADDITION AND SUBTRACTION

Adding whole numbers is combining two or more numbers (addends) to get a larger number called the **sum**, or **total**. Addition is indicated by the symbol "+", called the plus sign.

To **add** whole numbers, line up the digits according to their place value: ones digits under the ones digits, tens digits under the tens digits, etc. After the digits are lined up, add the ones digits. Next, add the tens digits, then the hundreds digits, etc. Place each sum below the line in the proper column.

Example A Find the sum of 305 and 241.

Line up the digits → Add the ones → Add the tens → Add the hundreds

$$
\begin{array}{r} 305 \\ +241 \\ \hline \end{array}
\qquad
\begin{array}{r} 305 \\ +241 \\ \hline 6 \end{array}
\qquad
\begin{array}{r} 305 \\ +241 \\ \hline 46 \end{array}
\qquad
\begin{array}{r} 305 \\ +241 \\ \hline 546 \end{array} \leftarrow \text{Answer}
$$

$$5 + 1 = 6 \qquad 0 + 4 = 4 \qquad 3 + 2 = 5$$

When the sum of the digits in any one place value column is ten or greater, it is necessary to show less, or carry, because only one digit can be written in a place value column. This can be done by regrouping.

Example B   Add the ones → Add the tens → Add the hundreds

$$
\begin{array}{r} {}^{1} \\ 437 \\ 672 \\ +195 \\ \hline 4 \end{array}
\qquad
\begin{array}{r} {}^{21} \\ 437 \\ 672 \\ +195 \\ \hline 04 \end{array}
\qquad
\begin{array}{r} {}^{21} \\ 437 \\ 672 \\ +195 \\ \hline 1304 \end{array} \leftarrow \text{Answer}
$$

$7 + 2 + 5 = 14$	$1 + 3 + 7 + 9 = 20$	$2 + 4 + 6 + 1 = 13$
Regroup.	Regroup.	Regroup.
14 ones = 1 ten	20 tens = 2 hundreds	13 hundreds = 1
plus 4 ones.	plus 0 tens.	thousand
Place the 1 ten	Place the 2 hundreds	plus 3 hundreds. Place the
in the tens column.	in the hundreds column.	1 thousand below the line.

Subtracting whole numbers is the opposite of adding. It is taking away a smaller number (subtrahend) from a larger number (minuend). The answer is the **difference**, or **remainder**. Subtraction is indicated by the symbol "−", called the minus sign.

To subtract whole numbers, line up the digits. Place the digits in the number to be subtracted (the smaller number) under the digits in the number from which they are to be subtracted (the larger number). After the digits have been lined up, subtract the digits in the ones column, then in the tens column, then in the hundreds column, and so on.

Example C   Subtract 42 from 362.

Subtract the ones → Subtract the tens → Subtract the hundreds

362	362	362
− 42	− 42	− 42
0	20	320 ← Answer

$2 - 2 = 0$ $6 - 4 = 2$ Think of 0 as being
in the hundreds place in
the number being
subtracted. $3 - 0 = 3$

To subtract two whole numbers, always subtract the smaller number from the larger. However, digits in the number to be subtracted may be greater than the corresponding digits from which they are to be subtracted. When this occurs, it is necessary to borrow, or show more by regrouping.

Sometimes it is necessary to regroup in more than one column in the same problem. First, where needed, regroup in the column farthest to the right. Then, move to the left, regrouping in the columns where needed.

Example D   Subtract 155 from 432.

Problem → Regroup to show more → Regroup to show more → Subtract in
ones. Then subtract in   tens. Then, subtract in   the hundreds
the ones column.   the tens column.   column.

	2 12	12 3 12	12 3 12
432	4̶3̶2	4̶3̶2	4̶3̶2
−155	−155	−155	−155
	7	77	277

3 tens 2 ones =   4 hundreds 2 tens = $3 - 1 = 2$
2 tens 12 ones   3 hundreds 12 tens
$12 - 5 = 7$ $12 - 5 = 7$

If the digit you must use to regroup, or borrow from, is 0, move to the left until you reach a digit that is not a 0 to begin regrouping.

Example E   Subtract 3297 from 8005.

Regroup to show → Regroup to show → Regroup to show more
more hundreds   more tens   ones and subtract

7 10	7 9 10 10	7 9 10 9 10 15	7 9 10 9 10 15
8̶ 0̶ 0 5	8̶ 0̶ 0̶ 5	8̶ 0̶ 0̶ 5̶	8̶ 0̶ 0̶ 5̶
− 3 2 9 7	− 3 2 9 7	− 3 2 9 7	− 3 2 9 7
			4 7 0 8

ERIC BELLAMY

NAME

23R   09/26/88

SECTION   DATE

INSTRUCTOR

A1.2 PROBLEMS

Add.

1. $165 + 387$

   552

2. $492 + 278$

   770

3. $556 + 177$

   733

4. $631 + 582$

   1213

5. $784 + 865$

   1649

6. $931 + 687$

   1618

7. $275 + 168$

   443

8. $382 + 449$

   831

9. $546 + 275$

   821

10. 329
    $+296$
    625

11. 187
    $+655$
    842

12. 489
    $+421$
    910

13. 715
    $+296$
    1011

14. 834
    $+189$
    1023

15. 958
    $+ 93$
    1051

16. 657
    $+ 84$
    741

17. 865
     176
    $+475$
    1516

18. 327
     885
    $+193$
    1405

19. 254
     482
    $+816$
    1552

20. 637
     184
    $+548$
    1369

21. 265
     384
    $+627$
    1276

22. 930
     846
    $+175$
    1951

23. 1648
     2771
    $+6935$
    11,354

24. 3008
     4962
    $+7517$
    15,487

25. 6644
     2288
    $+6623$
    15,555

26. 9339
     2772
    $+7889$
    20,000

27. 7654
     4567
    $+1110$
    13,331

Subtract.

28. 126 − 64

29. 284 − 194

30. 653 − 573

31. 133 − 55

32. 228 − 49

33. 437 − 178

34.
```
  354
− 168
  186
```

35.
```
  846
− 569
  277
```

36.
```
  961
− 472
  489
```

37.
```
  8271
− 6395
  1876
```

38.
```
  5726
− 2879
  2847
```

39.
```
  6437
− 5682
   755
```

40.
```
  7825
− 5937
  1888
```

41.
```
  9234
− 6345
  2889
```

42.
```
  4321
− 1234
  3087
```

43.
```
  7641
− 1467
  6174
```

44.
```
  8532
− 2358
  6174
```

45.
```
  9753
− 3579
  6174
```

46. 2026 − 178
```
   178
  1848
```

47. 4065 − 389
```
   389
  3676
```

48. 7058 − 669
```
   669
  6389
```

49. 5000 − 557
```
   557
  4443
```

50. 3000 − 2178
```
  2178
   822
```

51. 8053 − 7668
```
  7668
   385
```

52.
```
  4005
− 3066
   939
```

53.
```
  8002
− 7063
   939
```

54.
```
  2047
− 1767
   280
```

A1.3 MULTIPLICATION AND DIVISION

Multiplication is a short-cut method of doing addition. The numbers being multiplied are called **factors**. The answer to a multiplication problem is called the **product**. Multiplication is indicated by an "×" sign, a dot "·", or by parentheses "( ) ( )". To multiply, align the digits in the numbers being multiplied according to their place value. Then multiply as shown.

Example A   Find 12 × 43.

Think of 43 as being multiplied by 1 ten and 2 ones.

Problem → Multiply 2 → Multiply 1 → Add partial products
Align digits   ones times 43   tens times 43
according to
place value

$$
\begin{array}{r} 43 \\ \times 12 \\ \hline \end{array}
\qquad
\begin{array}{r} 43 \\ \times 12 \\ \hline 86 \end{array}
\qquad
\begin{array}{r} 43 \\ \times 12 \\ \hline 86 \\ 430 \end{array}
\qquad
\begin{array}{r} 43 \\ \times 12 \\ \hline 86 \\ 430 \\ \hline 516 \end{array} \leftarrow \text{Answer}
$$

    2 × 43   1 ten × 43   86 and 430 are
    = 86   = 43 tens   partial products
                    or 430

Sometimes when you multiply one digit by another, you get a product of 10 or greater. When this occurs, regroup to name less in the column. The amount taken out of the given column is added in the column to the left.

Example B   Find (4)(346).

Problem → Multiply 4 × 6 → Multiply 4 × 4 → Multiply 4 × 3

$$
\begin{array}{r} 346 \\ \times\ \ 4 \\ \hline \end{array}
\qquad
\begin{array}{r} {}^{2} \\ 346 \\ \times\ \ 4 \\ \hline 4 \end{array}
\qquad
\begin{array}{r} {}^{1\,2} \\ 346 \\ \times\ \ 4 \\ \hline 84 \end{array}
\qquad
\begin{array}{r} {}^{1\,2} \\ 346 \\ \times\ \ 4 \\ \hline 1384 \end{array} \leftarrow \text{Answer}
$$

4 × 6 = 24   4 × 4 = 16   4 × 3 = 12
Put 4 in the ones   16 + 2 = 18   12 + 1 = 13
place in the answer. Put 8 in tens   Put 13 in answer.
Put 2 over the   place in answer.
number in the tens   Put 1 over number
place.   in hundred place.

It is not necessary to multiply a 0. Instead, put 0's in the appropriate place value columns in the partial products as place holders.

11

Example C Find 104 × 356.

$$
\begin{array}{r}
356 \\
\times\ 104 \\
\hline
1424 \\
35600 \\
\hline
37024 \\
\end{array}
$$

356
× 104 ← (1 hundred 0 tens 4 ones)
————
1424 ← (4 ones × 356)
(1 hundred × 356) → 35600 ← (0's as place holders for 0 tens × 356)
————
37024 ← (1424 + 35600)

Division is the inverse of multiplication. The number being divided is called the **dividend**. The number by which you are dividing is called the **divisor**. The answer to a division problem is called the **quotient**.

The symbols used to indicate division are: the sign "÷" with the dividend always written in front of the sign and the divisor after it, a bar "—" with the dividend on top of the bar and the divisor below it, and a box " ⌐ " with the dividend inside and the divisor outside in front.

$$
\text{Divisor} \overline{)\text{Dividend}}^{\text{Quotient}} \qquad \frac{\text{Dividend}}{\text{Divisor}} = \text{Quotient}
$$

In division, divide from left to right in the dividend. To begin, move as many digits to the right as necessary in the dividend to have a "new" dividend that is as large as or larger than the divisor. After you have placed the first digit in the quotient, there must be a digit in the quotient for every digit remaining in the dividend. When no division can take place, 0's are used as place holders.

Example D Find 2912 ÷ 14.

This division problem asks, "How many 14's are in 2912?"

Begin with the "new" dividend 29.

```
        208
  14)2912    How many 14's in 29? 2. Put 2 in quotient above 9.
   -28    ← Multiply: 2 × 14 = 28.
   ———
    112   ← Subtract: 29 − 28 = 1. Bring down the 1.
          How many 14's in 11? None. Put 0 in quotient above 1.
          Bring down the 2.
          How many 14's in 112? 8. Put 8 in quotient above 2.
   -112   ← Multiply: 8 × 14 = 112.
   ———
      0   ← Subtract: 112 − 112 = 0.
```

The quotient is 208. There is no remainder.

When division does not "come out even" the quotient has a remainder.

Example E Find 892 ÷ 15.

```
        59
  15)892    How many 15's in 89? 5      The answer can be written
   -75   ← 5 × 15                          either as 59r7 or as
   ———
    142    How many 15's in 142? 9      59 7/15  ← Remainder
   -135  ← 9 × 15                              ← Divisor
   ———
      7  ← Remainder                     The remainder must be
                                         less than the divisor.
```

ERIC BELLAMY

NAME

* WENT ON TO
ACHIEVEMENT TEST

SECTION   DATE   INSTRUCTOR

A1.3 PROBLEMS

Multiply.

1. $\begin{array}{r} 64 \\ \times 10 \end{array}$	2. $\begin{array}{r} 25 \\ \times 34 \end{array}$	3. $\begin{array}{r} 36 \\ \times 19 \end{array}$
4. $\begin{array}{r} 321 \\ \times\ 42 \end{array}$	5. $\begin{array}{r} 457 \\ \times\ 64 \end{array}$	6. $\begin{array}{r} 821 \\ \times\ 71 \end{array}$
7. $\begin{array}{r} 920 \\ \times\ 42 \end{array}$	8. $\begin{array}{r} 479 \\ \times\ 68 \end{array}$	9. $\begin{array}{r} 635 \\ \times\ 97 \end{array}$
10. $\begin{array}{r} 307 \\ \times\ 27 \end{array}$	11. $\begin{array}{r} 461 \\ \times\ 84 \end{array}$	12. $\begin{array}{r} 897 \\ \times\ 65 \end{array}$
13. $\begin{array}{r} 287 \\ \times\ 87 \end{array}$	14. $\begin{array}{r} 142 \\ \times\ 90 \end{array}$	15. $\begin{array}{r} 839 \\ \times\ 30 \end{array}$
16. $\begin{array}{r} 492 \\ \times\ 77 \end{array}$	17. $\begin{array}{r} 403 \\ \times 126 \end{array}$	18. $\begin{array}{r} 215 \\ \times 104 \end{array}$
19. $\begin{array}{r} 129 \\ \times 575 \end{array}$	20. $\begin{array}{r} 806 \\ \times 329 \end{array}$	21. $\begin{array}{r} 420 \\ \times 412 \end{array}$
22. $\begin{array}{r} 397 \\ \times 620 \end{array}$	23. $\begin{array}{r} 584 \\ \times 707 \end{array}$	24. $\begin{array}{r} 963 \\ \times 854 \end{array}$
25. $\begin{array}{r} 657 \\ \times 332 \end{array}$	26. $\begin{array}{r} 1101 \\ \times\ 321 \end{array}$	27. $\begin{array}{r} 1321 \\ \times\ 65 \end{array}$

Divide.

28. $14\overline{)182}$

29. $15\overline{)165}$

30. $23\overline{)483}$

31. $27\overline{)1053}$

32. $46\overline{)1196}$

33. $39\overline{)2223}$

34. $1887 \div 17$

35. $\dfrac{22924}{44}$

36. $\dfrac{32785}{83}$

37. $25\overline{)6230}$

38. $48\overline{)9720}$

39. $33\overline{)2874}$

40. $26\overline{)1439}$

41. $17\overline{)5121}$

42. $21\overline{)14810}$

43. $123\overline{)13994}$

44. $331\overline{)28007}$

45. $48\overline{)4009}$

46. $52\overline{)20009}$

47. $128\overline{)8329}$

48. $225\overline{)7660}$

A1.4 INTEGERS AND THEIR OPERATIONS

The number line includes numbers to the right and left of 0. It extends infinitely in both directions.

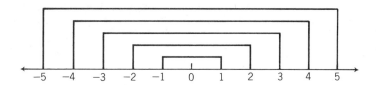

As you can see on the number line, the numbers $1, 2, 3, \ldots$ are to the right of 0. The numbers to the right of 0 are called **positive** and are sometimes written with a " $+$ " in front of them. The numbers to the left of 0 are called **negative** and are written with a " $-$ " sign in front of them. Zero is neither positive nor negative. The whole numbers—positive, negative, and 0—are called **integers**.

Every integer has an opposite. A nonzero integer and its opposite are the same distance from 0 on the number line but are in **opposite** directions. Zero is its own opposite. Opposites are shown on the number line above.

Example A

$+1$ is the opposite of -1	-5 is the opposite of $+5$
-3 is the opposite of $+3$	-6 is the opposite of $+6$
-17 is the opposite of 17	35 is the opposite of -35
162 is the opposite of -162	-3840 is the opposite of 3840

To add two integers, you can use the following rules.

> 1. If the addends have the same sign, add the two numbers and prefix their common sign:

Example B $(+62) + (+14) = +76.$

Example C $(-29) + (-13) = -42.$

> 2. If the addends have different signs, find the difference of the two numbers and prefix the sign of the number that is the greater distance from 0.

Example D $(15) + (-8) = +7.$

Example E $7 + (-17) = -10.$

To subtract two integers, rewrite the problem as a corresponding addition problem. That is, to subtract an integer, add its opposite.

Example F

Subtraction problem	Change to addition problem	Answer
$(+6) - (+2)$ $\rightarrow$	$(+6) + (-2)$ $=$	$+4$

Change sign to opposite

Example G   Subtraction problem   Change to addition problem   Answer

$$(+56) - (-37) \quad \rightarrow \quad (+56) + (+37) \quad = \quad +93$$

Change sign to opposite

To multipy two integers, you can use the following rules:

| 1. The product of two positive integers is positive.

Example H $(+4) \times (+2) = +8.$

| 2. The product of a positive integer and a negative integer is negative.

Example I $(+3) \times (-6) = -18.$

Example J $(-7) \times (+4) = -28.$

| 3. The product of two negative integers is positive.

Example K $(-3) \times (-2) = +6.$

To divide two integers, you can use the following rules:

| 1. When a positive integer is divided by a positive integer, the quotient is positive.

Example L $(+8) \div (+4) = +2.$

| 2. When a positive integer is divided by a negative integer, the quotient is negative.

Example M $(+10) \div (-2) = -5.$

| 3. When a negative integer is divided by a negative integer, the quotient is positive.

Example N $(-6) \div (-3) = +2.$

| 4. When a negative integer is divided by a positive integer, the quotient is negative.

Example O $(-18) \div (+6) = -3.$

A1.4 PROBLEMS   Add.

1. $(+3) + (+5)$ ___8___

2. $(+8) + (+2)$ ___10___

3. $(+5) + (-6)$ ___-1___

4. $(+9) + (-4)$ ___5___

5. $(-2) + (-8)$ ___-10___

6. $(-7) + (+1)$ ___-6___

7. $(-73) + (+47)$ ___-26___

8. $(+86) + (-59)$ ___25___

9. $(-78) + (-30)$ ___-108___

10. $(-100) + (+50)$ ___-50___

11. $(+75) + (-25)$ ___50___

12. $(+150) + (+50)$ ___200___

13. $(-200) + (+100)$ ___-100___

14. $(+132) + (-181)$ ___-49___

15. $(+355) + (-163)$ ___192___

16. $(-461) + (+401)$ ___-60___

17. $(-250) + (-175)$ ___-425___

18. $(+1000) + (+100)$ ___1100___

19. $(+1264) + (-1575)$ ___-311___

20. $(-3508) + (3467)$ ___-41___

Subtract.

21. $(+2) - (+7)$ ___-5___

22. $(-5) - (-6)$ ___1___

23. $(+7) - (-1)$ ___8___

24. $(-8) - (+1)$ ___-7___

25. $(+25) - (-15)$ ___40___

26. $(-33) - (-49)$ ___16___

27. $(-55) - (+11)$ ___44___

28. $(-70) - (-56)$ ___-14___

29. $(+71) - (+31)$ ___40___

30. $(+88) - (+19)$ ___69___

31. $(+62) - (-44)$ ___106___

32. $(+58) - (-48)$ ___106___

33. $(-39) - (-29)$ ___-10___

34. $(-72) - (-71)$ ___-1___

35. $(-150) - (+50)$ ___-200___

36. $(-200) - (+100)$ ___-100___

37. $(+123) - (+321)$ ___-198___

38. $(+109) - (+405)$ ___-296___

39. $(-325) - (-150)$ ___-175___

40. $(-175) - (-220)$ ___45___

17

Multiply.

41. $(+3) \times (+1)$ ___3___   42. $(+2) \times (+5)$ ___10___

43. $(+7) \times (+6)$ ___42___   44. $(+9) \times (+8)$ ___72___

45. $(+5) \times (-4)$ ___-20___   46. $(+3) \times (-8)$ ___-24___

47. $(+7) \times (-2)$ ___-14___   48. $(+4) \times (-6)$ ___-24___

49. $(-8) \times (-6)$ ___40___   50. $(-5) \times (-5)$ ___25___

51. $(-9) \times (-10)$ ___90___   52. $(-7) \times (-4)$ ___28___

53. $(-12) \times (+5)$ ___-60___   54. $(-11) \times (+4)$ ___-44___

55. $(-20) \times (+255)$ ___-5100___   56. $(-10) \times (+100)$ ___-1000___

57. $(+16) \times (0)$ ___0___   58. $(+15) \times 0$ ___0___

59. $(-30) \times (-3)$ ___90___   60. $(-40) \times (-4)$ ___160___

Divide.

61. $(+9) \div (+3)$ ___3___   62. $(+12) \div (+4)$ ___3___

63. $(+2) \div (+1)$ ___2___   64. $(+4) \div (+2)$ ___2___

65. $(+10) \div (-5)$ ___-2___   66. $(+12) \div (+6)$ ___2___

67. $(+14) \div (-2)$ ___-7___   68. $(+16) \div (-4)$ ___-4___

69. $(-12) \div (-3)$ ___4___   70. $(-14) \div (-7)$ ___2___

71. $(-10) \div (-2)$ ___5___   72. $(-18) \div (-6)$ ___3___

73. $(-25) \div (+5)$ ___-5___   74. $(-20) \div (+4)$ ___-5___

75. $(+25) \div (-5)$ ___-5___   76. $(-30) \div (+10)$ ___-3___

77. $(-45) \div (+15)$ ___-3___   78. $(-100) \div (-25)$ ___4___

79. $(+48) \div (+12)$ ___4___   80. $(+52) \div (+13)$ ___4___

A1.5 ORDER OF OPERATIONS AND GROUPING

When you have a series of operations to perform, and there are no parentheses present, first do the multiplications and divisions in order from left to right, then do the additions and subtractions in order from left to right.

Example A Find $3 \times 4 + 16 \div 2 - 3$.

$$12 \quad + \quad 8 \quad - 3$$
$$17$$

Example B Find $5 \ -2 + 4 \times 7$.

$$5 \ -2 + \quad 28$$
$$31$$

Example C Find $(-25) \div 5 - 5 + (-5 \times 4)$.

$$-5 \quad -5 + (-20)$$
$$-31$$

Example D Find $3 \times 6 + 8 \div 4 - 6 \div 3 + 3 \times 4 \div 2$.

$$18 \ + \ 2 \ - \ 2 \ + \ 12 \div 2$$
$$18 \ + \ 2 \ - \ 2 \ + \quad 6$$
$$24$$

Parentheses may be used to group numbers that are to be treated as a single quantity. When parentheses are present, perform the operations within the parentheses first. Then remove the parentheses and use the rules for the order of operations.

Example E Find $(-3) \times (5 + 8)$.

$$(5 + 8) = 13 \qquad (-3) \times 13 = -39$$

So, $(-3) \times (5 + 8) = -39$

Example F Find $(6 \times 3) \div (7 + 2)$.

$$(6 \times 3) = 18 \qquad (7 + 2) = 9 \qquad (18 \div 9) = 2$$

So, $(6 \times 3) \div (7 + 2) = 2$

Example G Find $(-18) + (15 - 3)$.

$$(15 - 3) = 12 \qquad (-18) + 12 = -6$$

So, $-18 + (15 - 3) = -6$

Example H Find $(36 - 14) - (23 - 5)$.

$$(36 - 14) = 22 \qquad (23 - 5) = 18 \qquad 22 - 18 = 4$$

So, $(36 - 14) - (23 - 5) = 4$

A number immediately preceding or following a parentheses, with no sign in between, means that you can either multiply every term within the parentheses by that number or perform the operation within the parentheses first and then multiply by the number on the outside.

Example I   Find $5(13 - 6)$.

This can be done in two ways.

$$5(13 - 6) = 5(13) - 5(6) = 65 - 30 = 35$$

or

$$5(13 - 6) = 5(7) = 35$$

Example J   Find $(24 + 16)(-2)$.

$$(24 + 16)(-2) = (24)(-2) + (16)(-2) = -48 - 32 = -80$$

Example K   Find $8(14 + 2) - 6(5 - 10)$.

$$8(14 + 2) - 6(5 - 10) = 8(14) + 8(2) - 6(5) - 6(-10)$$

$$= 112 + 16 - 30 + 60 = 158$$

or

$$8(14 + 2) - 6(5 - 10) = 8(16) - 6(-5) = 128 + 30 = 158.$$

Multiplying every term within the parentheses by the number immediately preceding or following it is called the **distributive property** and is very useful when solving certain types of algebra problems.

Brackets, "[ ]", may be used in the same way as parentheses. Sometimes more than one grouping symbol must be used in a problem, and you may have both parentheses and brackets. When a problem contains more than one grouping symbol, perform the operations within the innermost grouping symbols first. Then, work toward the outermost grouping symbols.

Example L   Find $7 + [5 + 3(18 \div 9)]$.

$$7 \ + [5 + 3(2)]$$
$$7 \ + \ [5 + 6]$$
$$7 \ + \quad [11]$$
$$18$$

Example M   Find $42 - [8 + (24 \div 6)3] - 3[7 - (10 \div 5)]$.

$$42 - [8 + (4)3] - 3[7 - 2]$$
$$42 - [8 + 12] \ - 3[7 - 2]$$
$$42 - \quad [20] \quad - \quad 3[5]$$
$$42 - \quad 20 \quad - \quad 15$$
$$7$$

A1.5 PROBLEMS

Solve.

1. $(3 \times 5) + (7 \times 8) =$ _____ 71

2. $4 - 9 + 6 \times 2 =$ _____ 7

3. $18 \div 9 + 2 \times 6 =$ _____ 14

4. $45 \div 9 \times (-5) =$ _____ -25

5. $(63 \div 21) \times 3 + 97 =$ _____ 106

6. $14 - 7 \times 6 \div 2 + 8 =$ _____ 29

7. $47 + 128 \div 32 \times (-9) =$ _____

8. $0 + 167 \times 0 =$ _____

9. $63 \div 9 \times 7 =$ _____ 49

10. $72 \times 8 \div 9 =$ _____

11. $70 \div 7 + 70 \div 10 =$ _____ 17

12. $70 \div 7 - 70 \div 10 =$ _____

13. $70 \div 10 + 70 \div 7 =$ _____

14. $70 \div 10 - 70 \div 7 =$ _____

15. $15 \div 5 \times (-3) + 9 =$ _____ 0

16. $5 \times 3 \div (-15) + 9 =$ _____

17. $(15 \div 3) \times 5 + 9 =$ _____

18. $26 \div 13 \times 2 - 17 =$ _____

19. $72 \div 8 \times 9 \div 9 =$ _____

20. $48 \div 16 \times 4 \div 3 =$ _____

21. $1296 \div 36 + 6 \times (-9) - 36 =$ _____

22. $52 \div 13 \times 3 \div 2 =$ _____

23. $1728 \div 48 \div 36 \times 9 \div 3 =$ _____

24. $50 \div 2 + 25 \div 5 - 5 =$ _____

25. $(-52) \div (-26) \div 2 \times (-13) =$ _____

26. $36 + (-40) \times 10 \div 4 - 12 =$ _____

27. $12 \div (-4) + (-3) \times (-6) \div 2 - 18 =$ _____

28. $48 + (-36) \div (-9) \times (-4) - 16 =$ _____

29. $5 + (4 - 3)2 =$ _____

30. $7 + (8 - 5)3 =$ _____

31. $5(16 \div 2) - 30 =$ _____

32. $9(14 \div 7) - 18 =$ _____

33. $3(5 + 6) =$ _____

34. $9(8 + 3) =$ _____

35. $17 - (7 \div 7) =$ _____

36. $18 - (18 \div 18) =$ _____

37. $36 + (36 \div 9) =$ _____

38. $81 + (9 \times 9) =$ _____

39. $(9 + 21) \div 5 =$ _____

40. $6 + (2 \times 13) =$ _____

41. $50 \div (60 - 10) =$ _____

42. $17 - (18 \div 18) =$ _____

43. $(6 + 7) \times (60 \div 15) =$ _____

44. $(4 - 4) \cdot (4 \div 4) =$ _____

45. $15[3 + (8 - 2)5] =$ _____

46. $85 - 5[5(8 - 5)] =$ _____

47. $24 - [1 + (9 \div 9)3] =$ _____

48. $18 - [2 + (7 \div 7)2] =$ _____

49. $6[6 + 8(4 + 2)] =$ _____

50. $12 \div [(15 - 3) \div 1] =$ _____

51. $3 + [17 - (17 \div 17)] + 4(16 \div 4) =$ _____

52. $8 + [12 - (15 \div 5)3] + 6(12 \div 4)2 =$ _____

53. $9[(36 \div 9) \div 4] - (9 \div 9) =$ _____

54. $6[(24 \div 4) - 6] =$ _____

55. $12 - [3 \times (12 \div 3)] =$ _____

56. $18 - [(21 \div 7)3] =$ _____

57. $15 + 3[5 + (15 \div 3)5] - [7(7 \div 7)] =$ _____

58. $13 - 2[7 + (18 \div 6)2] - [8(8 \div 8)] =$ _____

59. $39 + 6[35 - 3(14 \div 7) - 29] =$ _____

60. $47 + 5[15 - 3(15 \div 5) - 6] =$ _____

A1.6 INEQUALITIES

Numbers are used to tell "how many" and to give the "order" or position of events or objects. The number 3 can mean that there are 3 objects, or it can mean that an object is the third one in line. Three symbols are used for describing how numbers are ordered. They are:

= Equals   > Greater than   < Less than

The symbols " < " and " > " are used to show the relationship between two numbers that are *not* equal to each other.

A visualization of these relationships is possible by using a number line. On it, "greater than" numbers are to the right of "less than" numbers.

Example A

2 is "less than" 5 $2 < 5$
6 is "greater than" 4 $6 > 4$
7 "equals" 7 $7 = 7$

Example B $-5 < -2.$

Example C $9{,}584 < 9{,}585.$

"Less than" and "greater than" relationships may be used to denote more than one number. (The letter "x" will be used to represent a number or numbers.)

Example D   Using only the whole numbers, 0, 1, 2, etc., what numbers are less than five?

$x < 5$ 0, 1, 2, 3, and 4 are less than 5

Example E   What whole numbers are greater than 7?

$x > 7$ 8, 9, 10, 11, etc., are greater than 7

The relationship "equals" may be used in conjunction with either the greater than or less than relationship. When equals is used with either of these two, the word "or" is used in reading the expression.

Example F $x \leq 6.$

If x is a whole number, this is read:

x is a whole number that is less than or equal to 6. The numbers are: 0, 1, 2, 3, 4, 5, and 6.

Example G $x \geq 3$.

If x is a whole number, this is read:

x is greater than or equal to 3. The numbers are: 3, 4, 5, 6, 7, etc.

The relationships less than and greater than may also be used to denote an interval, or a set of numbers, between the larger and the smaller number.

Example H $-3 < x < 9$.

If x is an integer, this is read:

-3 is less than x and x is less than 9,

or it is read:

x is greater than -3 and less than 9. In either case, the integers that satisfy both conditions are: -2, -1, 0, 1, 2, 3, 4, 5, 6, 7, and 8.

Example I $0 < x < 5$.

If x is an integer, the numbers 1, 2, 3, and 4 are greater than 0 and less than 5.

The three relationships " $=$ " " $<$," and " $>$ " may be used jointly in many different situations.

Example J $4 < x \leq 9$.

If x is an integer, 5, 6, 7, 8, and 9 are greater than 4 and less than or equal to 9.

Example K $5 \leq x < 7$.

If x is an integer, 5 and 6 are the integers that are greater than or equal to 5 and less than 7.

Example L $-2 \leq x \leq 6$.

If x is an integer, -2, -1, 0, 1, 2, 3, 4, 5, and 6 are the integers that are greater than or equal to -2 and less than or equal to 6.

A1.6 PROBLEMS

Insert the proper symbol ($<$, $=$, or $>$) in each of the following blanks:

1. 5 $\leq$ 8

2. 9 $>$ $-$10

3. 0 $\leq$ 5

4. 7 $>$ 3

5. 103 $<$ 809

6. 5 $=$ 5

7. $-$35 $<$ 15

8. 45 $>$ $-$39

9. 74 $=$ 74

10. 503 $>$ 501

Write the proper words for the indicated relationship between the two numbers.

11. five ___is less than___ eight

12. nine ___is greater than___ seven

13. twelve ___is less than___ nineteen

14. zero ___is less than___ two

15. twenty ___is more than___ seven

What integers can be put in place of x to make these statements true?

16. $x < 5$ $x =$ 4, 3, 2, 1 0 -1 -2 . . .

17. $x < -8$ $x =$ -9, -10, -11, . . .

18. $x > 7$ $x =$ 8, 9, 10, . . .

19. $x > -6$ $x =$ -5, -4, -3, . . .

20. $x \leq 3$ $x =$ 3, 2, 1, 0, -1 . . .

21. $x \leq -7$ $x =$ -7, -8, -9 . . .

22. $3 \leq x < 6$ $x =$ 3, 4, 5

23. $0 < x \leq 2$ $x =$ 1, 2

25

24. $-4 \leq x \leq 9$ $x =$ _4, 3 → 9_

25. $5 < x \leq 6$ $x =$ _6_

26. $201 \leq x \leq 506$ $x =$ _201 → 506_

27. $336 \leq x \leq 340$ $x =$ _336, 332, 338, 339, 340_

Order the numbers from least to greatest.

28. $5, 7, 2, 9, 3$ _2, 3, 5, 7, 9_

29. $3, -4, -2, 0, 4$ _-4, -2, 0, 3, 4_

30. $101, 89, 102, 34, 29$ _29, 34, 89, 101, 102_

31. $19, 9, -7, 23, -20$ _-20, -7, 9, 19, 23_

In Problems 32 and 33, order the numbers from greatest to least.

32. $50, 39, 67, 42, 70, 31$ _70, 67, 50, 42, 39, 31_

33. $-49, 50, 101, -16, 121$ _121, 101, 50, -16, -49_

In Problems 34–41, list as indicated, or indicate that there is none.

34. Five integers less than 9. _8, 7, 6, 5, 4_

35. Three integers greater than -3. _-4, -5, -6_

36. All the whole numbers less than or equal to 4. _4, 3, 2, 1, 0_

37. The whole numbers greater than 1 and less than or equal to 5. _2, 3, 4, 5_

38. Whole numbers greater than or equal to -3 but less than 5. _0, 1, 2, 3, 4_

39. The whole numbers less than or equal to 6. _6, 5, 4, 3, 2, 1, 0_

40. The whole numbers greater than 1 and less than 2. _NONE_

41. The whole numbers less than 5 and greater than 4. _NONE_

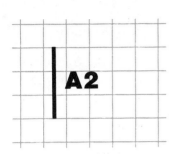

A2

FRACTIONS AND THEIR OPERATIONS

A2.1 MEANING OF FRACTIONS

Fractions are useful in expressing parts of things or quantities. An example is $\frac{3}{4}$, pronounced "three-fourths," meaning 3 of 4 equal parts. A fraction has a number below the line called the **denominator**, which indicates the size of the units $\left(\text{"fourths" in } \frac{3}{4}\right)$, and a number above the line called the **numerator**, which indicates how many units are involved $\left(\text{"three" in } \frac{3}{4}\right)$.

In a fraction, the denominator cannot be the number 0. The numerator of a fraction can be 0, but *the denominator of a fraction cannot be* 0.

Example A   Give $\frac{3}{4}$ of a day in terms of hours.

A day is cut into 4 equal parts by making each part 6 hours. Three of these equal parts is 18 hours. So $\frac{3}{4}$ of a day is 18 hours.

Example B   Give $\frac{11}{8}$ of a day in terms of hours.

A day has 8 equal parts of 3 hours each. 11 of those equal parts is 33 hours. So $\frac{11}{8}$ of a day is 33 hours.

Example C   Show $\frac{3}{4}$ on the number line.

On a line, plot 0 and 1, then cut this segment into 4 equal parts, as shown.

Move along 3 of the 4 equal parts to the right of 0. That point is $\frac{3}{4}$.

If the numerator of a fraction is greater then the denominator, the fraction is called an **improper fraction**. Every improper fraction can be written as a whole number or a **mixed number**.

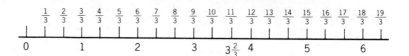

You can see that $\frac{11}{3}$ is $\frac{2}{3}$ more than 3 on the number line. We write $\frac{11}{3} = 3\frac{2}{3}$ (pronounced "three and two-thirds"). Also, you can see that

$$\frac{4}{3} = 1\frac{1}{3} \qquad \frac{5}{3} = 1\frac{2}{3} \qquad \frac{6}{3} = 2 \qquad \frac{7}{3} = 2\frac{1}{3}$$

$$\frac{8}{3} = 2\frac{2}{3} \qquad \frac{9}{3} = 3 \qquad \frac{10}{3} = 3\frac{1}{3}$$

and that

$$\frac{0}{3} = 0 \quad \text{and} \quad \frac{3}{3} = 1$$

Mixed numbers arise most often as quotients of whole numbers. Think of $\frac{14}{3}$ as 14 divided by 3.

$$\begin{array}{r} 4 \\ 3\overline{)14} \\ \underline{12} \\ 2 \end{array} \qquad \frac{14}{3} = 4\frac{2}{3}$$

The quotient is 4 and the remainder is 2.

Example D   Write $\frac{55}{6}$ as a mixed number.

$$\begin{array}{r} 9 \\ 6\overline{)55} \\ \underline{54} \\ 1 \end{array} \quad \begin{array}{l} \text{Quotient} = 9 \\ \\ \text{Remainder} = 1 \end{array} \qquad \text{Thus, } \frac{55}{6} = 9\frac{1}{6}.$$

Changing a mixed number into a fraction involves multiplication and addition.

Example E   Write $5\frac{4}{7}$ as an improper fraction.

The denominator will be 7. To find the numerator:

Multiply 5 and 7.   35
Then add 4.   39

$$5\frac{4}{7} = \frac{39}{7}$$

A2.1 PROBLEMS

Tell how many hours are represented.

1. $\frac{1}{6}$ of a day __4__

2. $\frac{5}{6}$ of a day __20__

3. $\frac{3}{8}$ of a day __9__

4. $\frac{7}{8}$ of a day __21__

5. $\frac{7}{12}$ of a day __14__

6. $\frac{4}{3}$ of a day __32__

Tell how many minutes are represented.

7. $\frac{5}{12}$ of an hour __25__

8. $\frac{9}{20}$ of an hour __27__

9. $\frac{13}{20}$ of an hour __39__

10. $\frac{17}{30}$ of an hour __34__

11. $\frac{11}{6}$ of an hour __110__

12. $\frac{13}{15}$ of an hour __52__

Express as a fraction of an hour (use a mixed number where appropriate):

13. 22 minutes __$\frac{11}{30}$__

14. 38 minutes __$\frac{19}{30}$__

15. 123 minutes __$2\frac{1}{20}$__

16. 207 minutes _____

Express as a fraction of a day (use a mixed number where appropriate):

17. 7 hours __$\frac{7}{24}$__

18. 42 hours __1 DAY 18 HRS__

19. 480 minutes __3 HRS__

20. 720 minutes __22 HRS__

Tell what fraction is marked by X on the number line.

21.

22.

23.

24.

$\frac{12\,3}{60}$

Write each improper fraction as a mixed number or a whole number.

25. $\dfrac{9}{4}$ = _____

26. $\dfrac{8}{3}$ = _____

27. $\dfrac{27}{12}$ = _____

28. $\dfrac{23}{10}$ = _____

29. $\dfrac{33}{21}$ = _____

30. $\dfrac{32}{18}$ = _____

31. $\dfrac{93}{31}$ = _____

32. $\dfrac{186}{62}$ = _____

33. $\dfrac{222}{74}$ = _____

34. $\dfrac{82}{28}$ = _____

35. $\dfrac{54}{45}$ = _____

36. $\dfrac{63}{36}$ = _____

37. $\dfrac{98}{28}$ = _____

38. $\dfrac{91}{26}$ = _____

39. $\dfrac{98}{21}$ = _____

40. $\dfrac{221}{55}$ = _____

41. $\dfrac{187}{38}$ = _____

42. $\dfrac{899}{93}$ = _____

Write each mixed number as an improper fraction.

43. $1\dfrac{3}{8}$ = _____

44. $2\dfrac{4}{9}$ = _____

45. $3\dfrac{10}{23}$ = _____

46. $4\dfrac{4}{9}$ = _____

47. $5\dfrac{9}{16}$ = _____

48. $6\dfrac{7}{11}$ = _____

49. $7\dfrac{2}{5}$ = _____

50. $8\dfrac{2}{3}$ = _____

51. $9\dfrac{1}{3}$ = _____

52. $10\dfrac{7}{8}$ = _____

53. $11\dfrac{5}{9}$ = _____

54. $12\dfrac{6}{7}$ = _____

55. $13\dfrac{4}{9}$ = _____

56. $14\dfrac{5}{8}$ = _____

57. $15\dfrac{9}{10}$ = _____

58. $16\dfrac{7}{10}$ = _____

59. $17\dfrac{11}{14}$ = _____

60. $18\dfrac{11}{13}$ = _____

61. $19\dfrac{9}{11}$ = _____

62. $20\dfrac{7}{12}$ = _____

63. $21\dfrac{13}{15}$ = _____

64. $22\dfrac{14}{17}$ = _____

65. $23\dfrac{11}{15}$ = _____

66. $24\dfrac{13}{16}$ = _____

67. $25\dfrac{15}{17}$ = _____

68. $26\dfrac{12}{13}$ = _____

69. $27\dfrac{17}{20}$ = _____

70. $28\dfrac{19}{30}$ = _____

71. $29\dfrac{21}{25}$ = _____

72. $30\dfrac{23}{26}$ = _____

A2.2 MULTIPLICATION AND DIVISION

Multiplication: To multiply fractions, multiply the numerators, then multiply the denominators.

Example A Compute $\frac{2}{3} \times \frac{5}{8}$.

$\frac{2}{3} \times \frac{5}{8} = \frac{2 \times 5}{3 \times 8} = \frac{10}{24}$. The fraction $\frac{10}{24}$ can be reduced to $\frac{5}{12}$ by dividing the numerator and denominator by 2.

Example B Compute $\frac{2}{3} \times \frac{6}{7} \times \frac{5}{6}$.

$$\frac{2}{3} \times \frac{6}{7} \times \frac{5}{6} = \frac{2 \times 6 \times 5}{3 \times 7 \times 6} = \frac{60}{126} = \frac{10}{21} \qquad \left(\frac{60 \div 6}{126 \div 6} = \frac{10}{21} \right)$$

Example C Finding the product $\frac{2}{3} \times \frac{6}{7} \times \frac{5}{6}$ could have been calculated more readily by what is called **cancelling**, that is, by dividing both the numerator and the denominator by 6 at the outset.

$$\frac{2}{3} \times \frac{\overset{1}{\cancel{6}}}{7} \times \frac{5}{\underset{1}{\cancel{6}}} = \frac{2 \times 5}{3 \times 7} = \frac{10}{21}$$

Example D Calculate $3\frac{3}{4} \times 1\frac{3}{5}$.

Convert the mixed numbers to fractions, then multiply the fractions. Cancel where possible.

$$3\frac{3}{4} \times 1\frac{3}{5} = \frac{\overset{3}{\cancel{15}}}{\underset{1}{\cancel{4}}} \times \frac{\overset{2}{\cancel{8}}}{\underset{1}{\cancel{5}}} = \frac{3 \times 2}{1 \times 1} = \frac{6}{1} = 6$$

Example E Find the product $2\frac{5}{8} \times 1\frac{3}{7} \times 3\frac{2}{5}$.

Again, convert the mixed numbers to fractions, and cancel.

$$2\frac{5}{8} \times 1\frac{3}{7} \times 3\frac{2}{5} = \frac{\overset{3}{\cancel{21}}}{\underset{4}{\cancel{8}}} \times \frac{\overset{2}{\cancel{10}}}{\underset{1}{\cancel{7}}} \times \frac{17}{\underset{1}{\cancel{5}}} = \frac{51}{4} = 12\frac{3}{4}$$

Make sure you see the cancelling that takes place to obtain numerator 51 and denominator 4.

Division: To divide fractions, invert the divisor and multiply.

Example F Calculate $\dfrac{5}{3} \div \dfrac{4}{9}$.

The divisor is $\dfrac{4}{9}$. The fraction inverted becomes $\dfrac{9}{4}$.

$$\frac{5}{3} \div \frac{4}{9} = \frac{5}{3} \times \frac{9}{4} = \frac{45}{12} = \frac{15}{4}$$

Example G Calculate $2\dfrac{1}{3} \div 4\dfrac{3}{5}$.

As before, convert to fractions, then divide.

$$2\frac{1}{3} \div 4\frac{3}{5} = \frac{7}{3} \div \frac{23}{5} = \frac{7}{3} \times \frac{5}{23} = \frac{35}{69}$$

Example H Calculate $-4\dfrac{3}{5} \div 2\dfrac{1}{3}$.

$$-4\frac{3}{5} \div 2\frac{1}{3} = \frac{-23}{5} \div \frac{7}{3} = \frac{-23}{5} \times \frac{3}{7} = \frac{-69}{35} = -1\frac{34}{35}$$

Example I Calculate $2\dfrac{1}{2} \div (-2)$.

$$2\frac{1}{2} \div (-2) = \frac{5}{2} \div \frac{-2}{1} = \frac{5}{2} \times \frac{-1}{2} = \frac{-5}{4} = -1\frac{1}{4}$$

Example J Calculate $2\dfrac{1}{2} \div \dfrac{1}{2}$.

$$2\frac{1}{2} \div \frac{1}{2} = \frac{5}{2} \div \frac{1}{2} = \frac{5}{\overset{}{\underset{1}{2}}} \times \frac{\overset{1}{2}}{1} = \frac{5}{1} = 5$$

Example K Calculate $\dfrac{3}{4} \div 1\dfrac{1}{8}$.

$$\frac{3}{4} \div 1\frac{1}{8} = \frac{3}{4} \div \frac{9}{8} = \frac{\overset{1}{3}}{\underset{1}{4}} \times \frac{\overset{2}{8}}{\underset{3}{9}} = \frac{2}{3}$$

Example L Calculate $15 \div \dfrac{3}{4}$.

$$15 \div \frac{3}{4} = \frac{\overset{5}{15}}{1} \times \frac{4}{\underset{1}{3}} = 20$$

A2.2 PROBLEMS

Multiply. Simplify where possible.

1. $\dfrac{3}{5} \times \dfrac{3}{4} =$ _____

2. $\dfrac{3}{8} \times \dfrac{1}{2} =$ _____

3. $\dfrac{5}{8} \times \dfrac{1}{4} =$ _____

4. $\dfrac{4}{5} \times \dfrac{2}{3} =$ _____

5. $\dfrac{4}{3} \times \dfrac{5}{9} =$ _____

6. $\dfrac{2}{5} \times \dfrac{4}{3} =$ _____

7. $\dfrac{3}{5} \times \dfrac{2}{5} =$ _____

8. $\dfrac{1}{3} \times \dfrac{3}{4} =$ _____

9. $\dfrac{1}{4} \times \dfrac{4}{5} =$ _____

10. $\dfrac{2}{3} \times \dfrac{3}{5} =$ _____

11. $\dfrac{4}{5} \times \dfrac{3}{4} =$ _____

12. $\dfrac{1}{6} \times \dfrac{3}{4} =$ _____

13. $\dfrac{2}{3} \times \dfrac{1}{6} =$ _____

14. $\dfrac{4}{5} \times \dfrac{1}{2} =$ _____

15. $\dfrac{6}{7} \times \dfrac{1}{3} =$ _____

16. $\dfrac{3}{4} \times \dfrac{2}{5} =$ _____

17. $\dfrac{5}{6} \times \dfrac{2}{3} =$ _____

18. $\dfrac{5}{8} \times \dfrac{7}{10} =$ _____

19. $\dfrac{2}{3} \times \dfrac{-7}{8} =$ _____

20. $\dfrac{3}{4} \times \left(-\dfrac{5}{6}\right) =$ _____

21. $\dfrac{-4}{9} \times \dfrac{5}{8} =$ _____

22. $\dfrac{3}{8} \times \dfrac{2}{9} =$ _____

23. $\dfrac{3}{10} \times \dfrac{5}{6} =$ _____

24. $\dfrac{3}{4} \times \dfrac{2}{3} =$ _____

25. $2\dfrac{1}{2} \times 1\dfrac{1}{4} =$ _____

26. $2\dfrac{1}{3} \times 1\dfrac{1}{2} =$ _____

27. $3\dfrac{1}{2} \times 6 =$ _____

28. $2\dfrac{2}{3} \times 9 =$ _____

29. $1\dfrac{1}{5} \times 4\dfrac{1}{6} =$ _____

30. $2\dfrac{3}{10} \times 1\dfrac{2}{3} =$ _____

31. $3\dfrac{1}{2} \times 1\dfrac{1}{7} =$ _____

32. $2\dfrac{2}{3} \times 1\dfrac{1}{2} =$ _____

33. $4 \times 3\dfrac{1}{2} =$ _____

34. $6 \times 4\dfrac{2}{3} =$ _____

35. $1\dfrac{3}{5} \times 3\dfrac{3}{4} =$ _____

36. $2\dfrac{5}{8} \times 4\dfrac{4}{7} =$ _____

37. $5\dfrac{5}{9} \times 1\dfrac{4}{5} =$ _____

38. $6\dfrac{3}{7} \times 1\dfrac{5}{9} =$ _____

39. $\dfrac{1}{2} \times 12 \times \dfrac{2}{3} =$ _____

40. $\dfrac{3}{4} \times 24 \times \dfrac{5}{6} =$ _____

41. $2 \times 2\dfrac{1}{2} \times 1\dfrac{1}{4} =$ _____

42. $3 \times 2\dfrac{1}{3} \times 1\dfrac{1}{2} =$ _____

43. $1\dfrac{3}{7} \times 3\dfrac{2}{5} \times 2\dfrac{5}{8} =$ _____

44. $1\dfrac{1}{6} \times 2\dfrac{4}{7} \times 3\dfrac{3}{5} =$ _____

45. $1\dfrac{1}{2} \times 2\dfrac{2}{3} \times 3\dfrac{3}{4} =$ _____

Divide. Simplify where possible.

46. $\dfrac{2}{3} \div \dfrac{3}{4} =$ _____

47. $\dfrac{3}{4} \div \dfrac{2}{3} =$ _____

48. $\dfrac{3}{5} \div \dfrac{2}{7} =$ _____

49. $\dfrac{2}{7} \div \dfrac{3}{5} =$ _____

50. $\dfrac{2}{3} \div \dfrac{2}{5} =$ _____

51. $\dfrac{2}{5} \div \dfrac{2}{3} =$ _____

52. $\dfrac{6}{35} \div \dfrac{-3}{5} =$ _____

53. $\dfrac{10}{21} \div \dfrac{-5}{7} =$ _____

54. $\dfrac{14}{15} \div \dfrac{-7}{3} =$ _____

55. $\dfrac{21}{22} \div \dfrac{7}{2} =$ _____

56. $\dfrac{77}{30} \div \dfrac{7}{6} =$ _____

57. $\dfrac{55}{42} \div \dfrac{11}{14} =$ _____

58. $2\dfrac{2}{3} \div \dfrac{2}{7} =$ _____

59. $3\dfrac{3}{4} \div \dfrac{3}{7} =$ _____

60. $6\dfrac{3}{4} \div 1\dfrac{1}{2} =$ _____

61. $6\dfrac{3}{8} \div 4\dfrac{1}{4} =$ _____

62. $4\dfrac{2}{7} \div 1\dfrac{2}{3} =$ _____

63. $6\dfrac{2}{5} \div 2\dfrac{2}{3} =$ _____

64. $1\dfrac{13}{42} \div 1\dfrac{5}{6} =$ _____

65. $\dfrac{2}{3} \div 1\dfrac{1}{3} =$ _____

66. $\dfrac{3}{4} \div 2\dfrac{1}{4} =$ _____

67. $(-2) \div \dfrac{3}{8} =$ _____

68. $3 \div \dfrac{4}{7} =$ _____

69. $\dfrac{3}{5} \div 2 =$ _____

70. $\dfrac{2}{3} \div (-3) =$ _____

71. $\dfrac{3}{7} \div 6 =$ _____

72. $\dfrac{4}{9} \div 8 =$ _____

73. $5\dfrac{3}{5} \div 1\dfrac{1}{3} =$ _____

74. $7\dfrac{2}{7} \div 1\dfrac{1}{2} =$ _____

75. $5\dfrac{5}{8} \div \dfrac{15}{16} =$ _____

76. $7\dfrac{7}{9} \div \dfrac{35}{18} =$ _____

77. $3\dfrac{11}{15} \div 2\dfrac{1}{2} =$ _____

78. $3\dfrac{13}{14} \div 2\dfrac{1}{2} =$ _____

79. $\dfrac{4}{5} \div 2 =$ _____

80. $\dfrac{6}{13} \div 3 =$ _____

81. $\dfrac{5}{8} \div 3 =$ _____

82. $-4\dfrac{2}{7} \div 2\dfrac{4}{5} =$ _____

83. $4\dfrac{6}{7} \div 1\dfrac{3}{14} =$ _____

84. $-1\dfrac{1}{2} \div 7\dfrac{2}{7} =$ _____

85. $1\dfrac{1}{3} \div 5\dfrac{3}{5} =$ _____

86. $\dfrac{1}{2} \div \dfrac{1}{4} =$ _____

87. $\dfrac{1}{3} \div \dfrac{1}{6} =$ _____

88. $\dfrac{15}{16} \div 5\dfrac{5}{8} =$ _____

89. $2\dfrac{2}{3} \div \dfrac{-1}{3} =$ _____

90. $4\dfrac{4}{5} \div \dfrac{-1}{5} =$ _____

A2.3 ADDITION AND SUBTRACTION

Adding and subtracting fractions is just like adding and subtracting quantities involving items, units of measure, units of weight, etc. Just think of the denominator as a unit.

Example A   What is $\dfrac{1}{5} + \dfrac{3}{5}$?

You can think of, "What is 1 apple plus 3 apples?" It is 4 apples. Or, "What is 1 inch plus 3 inches?" It is 4 inches. In like manner, the problem of this example asks, "What is 1 fifth plus 3 fifths?" It is 4 fifths, so

$$\frac{1}{5} + \frac{3}{5} = \frac{4}{5}$$

Example B   Calculate $\dfrac{9}{4} - \dfrac{3}{4}$.

Think, if you wish, "What is 9 gallons minus 3 gallons?" It is 6 gallons. Likewise, 9 fourths minus 3 fourths is 6 fourths.

$$\frac{9}{4} - \frac{3}{4} = \frac{6}{4} = \frac{3}{2}.$$

Example C   Find the sum of $\dfrac{4}{5}$ and $\dfrac{7}{10}$.

$$\frac{4}{5} = \frac{4 \times 2}{5 \times 2} = \frac{8}{10}$$

Therefore,

$$\frac{4}{5} + \frac{7}{10} = \frac{8}{10} + \frac{7}{10} = \frac{15}{10} = \frac{3}{2}$$

Example D   Calculate $\dfrac{9}{24} - \dfrac{1}{8}$.

$$\frac{9}{24} = \frac{9 \div 3}{24 \div 3} = \frac{3}{8}$$

Therefore,

$$\frac{9}{24} - \frac{1}{8} = \frac{3}{8} - \frac{1}{8} = \frac{2}{8} = \frac{1}{4}$$

Example E   Calculate $3\frac{3}{4} - 1\frac{5}{8}$.

$$3\frac{3}{4} = \frac{15}{4} = \frac{30}{8} \quad \text{and} \quad 1\frac{5}{8} = \frac{13}{8}$$

Therefore,

$$3\frac{3}{4} - 1\frac{5}{8} = \frac{30}{8} - \frac{13}{8} = \frac{17}{8} = 2\frac{1}{8}.$$

Notice that in adding or subtracting any fractions, the technique is to write all fractions so that they have the same denominator, called a *common denominator*. It is then a simple matter to add or subtract numerators. If mixed numbers are involved, begin by rewriting them as improper fractions.

Example F   Find the sum of $\frac{3}{8}$ and $\frac{9}{10}$.

This may seem like trying to add 3 oranges to 9 tomatoes. But it isn't. Both the eighths and the tenths can be changed to the same units (denominators). By multiplying 8 by 5, and multiplying 10 by 4, the same product, 40, is obtained. 40 will be the common denominator.

$$\frac{3}{8} = \frac{3 \times 5}{8 \times 5} = \frac{15}{40} \qquad \frac{9}{10} = \frac{9 \times 4}{10 \times 4} = \frac{36}{40}$$

Thus,

$$\frac{3}{8} + \frac{9}{10} = \frac{15}{40} + \frac{36}{40} = \frac{15 + 36}{40} = \frac{51}{40}.$$

Example G   Compute $\frac{5}{6} - \frac{1}{9}$.

Again, the idea is to rewrite both fractions so that they have the same denominator. $6 \times 3 = 18$ and $9 \times 2 = 18$. A common denominator of 18 will work.

$$\frac{5}{6} = \frac{5 \times 3}{6 \times 3} = \frac{15}{18} \qquad \frac{1}{9} = \frac{1 \times 2}{9 \times 2} = \frac{2}{18}$$

Thus, $\frac{5}{6} - \frac{1}{9} = \frac{15}{18} - \frac{2}{18} = \frac{13}{18}.$

A2.3 PROBLEMS

Solve.

1. $\dfrac{1}{8} + \dfrac{3}{8} = $ _____

2. $\dfrac{1}{7} + \dfrac{4}{7} = $ _____

3. $\dfrac{2}{5} + \dfrac{2}{5} = $ _____

4. $\dfrac{7}{12} - \dfrac{5}{12} = $ _____

5. $\dfrac{5}{8} - \dfrac{3}{8} = $ _____

6. $\dfrac{5}{6} - \dfrac{1}{6} = $ _____

7. $\dfrac{38}{21} - \dfrac{18}{21} = $ _____

8. $\dfrac{5}{8} + \dfrac{3}{8} - \dfrac{13}{8} = $ _____

9. $\dfrac{6}{7} + \dfrac{5}{7} - \dfrac{15}{7} = $ _____

10. $4\dfrac{7}{8} - \dfrac{5}{8} = $ _____

11. $3\dfrac{7}{12} - 4\dfrac{11}{12} = $ _____

12. $2\dfrac{9}{14} - 3\dfrac{13}{14} = $ _____

13. $\dfrac{13}{24} + \dfrac{19}{24} - \dfrac{31}{24} - \dfrac{47}{24} = $ _____

14. $\dfrac{29}{36} + \dfrac{17}{36} - \dfrac{53}{36} - \dfrac{41}{36} = $ _____

15. $5\dfrac{13}{18} - 7\dfrac{11}{18} + 3\dfrac{5}{18} - \dfrac{17}{18} = $ _____

16. $4\dfrac{11}{21} - 6\dfrac{13}{21} + 2\dfrac{8}{21} - \dfrac{13}{21} = $ _____

17. $5\dfrac{3}{5} - 2\dfrac{1}{5} = $ _____

18. $6\dfrac{4}{7} - 4\dfrac{2}{7} = $ _____

19. $8\dfrac{2}{9} - 3\dfrac{7}{9} = $ _____

20. $\dfrac{2}{3} + \dfrac{2}{9} = $ _____

21. $\dfrac{1}{4} + \dfrac{3}{8} = $ _____

22. $\dfrac{5}{6} - \dfrac{1}{2} = $ _____

23. $3\dfrac{3}{8} + \dfrac{1}{4} = $ _____

24. $4\dfrac{2}{9} + \dfrac{1}{3} = $ _____

25. $1\dfrac{5}{6} - \dfrac{2}{3} = $ _____

26. $2\dfrac{7}{8} - \dfrac{3}{4} = $ _____

27. $2\dfrac{4}{7} - 3\dfrac{9}{14} = $ _____

28. $3\dfrac{6}{11} - 5\dfrac{13}{22} = $ _____

29. $\dfrac{5}{8} - \dfrac{3}{20} = $ _____

30. $\dfrac{7}{18} - \dfrac{1}{12} = $ _____

31. $\dfrac{5}{12} - \dfrac{1}{8} = $ _____

32. $\dfrac{2}{21} - \dfrac{5}{6} = $ _____

33. $\dfrac{4}{21} - \dfrac{1}{9} = $ _____

34. $\dfrac{7}{15} - \dfrac{7}{10} = $ _____

35. $\dfrac{4}{5} + \dfrac{1}{2} = $ _____

36. $\dfrac{1}{3} + \dfrac{3}{4} = $ _____

37. $\dfrac{3}{5} + \dfrac{1}{2} = $ _____

38. $\dfrac{11}{15} + \dfrac{7}{18} - \dfrac{5}{9} = $ _____

39. $\dfrac{9}{10} + \dfrac{5}{12} - \dfrac{7}{15} = $ _____

40. $-\dfrac{3}{4} - \dfrac{5}{6} + \dfrac{1}{8} = $ _____

41. $\dfrac{3}{2} + \dfrac{7}{6} + \dfrac{9}{8} + \dfrac{-13}{12} = $ _____

42. $\dfrac{4}{3} + \dfrac{6}{5} + \dfrac{11}{10} + \dfrac{-16}{15} = $ _____

43. $\dfrac{5}{8} - \dfrac{3}{28} = $ _____

44. $\dfrac{7}{10} - \dfrac{1}{12} = $ _____

45. $\dfrac{5}{18} - \dfrac{1}{8} = $ _____

46. $\dfrac{2}{21} - \dfrac{5}{9} = $ _____

47. $\dfrac{4}{21} - \dfrac{1}{6} = $ _____

48. $\dfrac{7}{25} - \dfrac{7}{10} = $ _____

49. $-2\dfrac{1}{2} + 4\dfrac{3}{4} = $ _____

50. $-3\dfrac{1}{3} + 5\dfrac{5}{6} = $ _____

51. $1\dfrac{5}{9} - 2\dfrac{1}{6} = $ _____

52. $3\dfrac{2}{3} + 4\dfrac{2}{5} = $ _____

53. $2\dfrac{7}{8} - 3\dfrac{3}{4} = $ _____

54. $1\dfrac{8}{9} - 2\dfrac{2}{3} = $ _____

55. $3\dfrac{5}{12} - 1\dfrac{3}{8} = $ _____

56. $4\dfrac{11}{18} - 2\dfrac{7}{12} = $ _____

57. $6\dfrac{7}{10} - 2\dfrac{5}{6} = $ _____

58. $5\dfrac{5}{21} - 1\dfrac{5}{6} = $ _____

59. $1\dfrac{1}{2} + 2\dfrac{1}{3} + 3\dfrac{1}{6} = $ _____

60. $1\dfrac{2}{3} + 2\dfrac{1}{4} + 3\dfrac{1}{12} = $ _____

61. $8\dfrac{7}{10} - \dfrac{11}{12} - \dfrac{16}{15} = $ _____

62. $7\dfrac{13}{15} - \dfrac{17}{18} - \dfrac{11}{9} = $ _____

63. $-3\dfrac{3}{5} - 2\dfrac{5}{6} + 1\dfrac{9}{10} = $ _____

64. $-4\dfrac{3}{4} - 3\dfrac{1}{8} + 1\dfrac{5}{6} = $ _____

65. $2\dfrac{3}{22} - 1\dfrac{3}{4} - \left(\dfrac{-2}{3}\right) = $ _____

66. $3\dfrac{5}{21} - 2\dfrac{7}{9} - \left(\dfrac{-1}{2}\right) = $ _____

67. $4\dfrac{8}{15} + 5\dfrac{7}{9} + 3\dfrac{23}{45} + 2\dfrac{1}{3} = $ _____

68. $5\dfrac{7}{10} + 3\dfrac{6}{25} - 2\dfrac{13}{20} + 4\dfrac{1}{2} = $ _____

69. $7\dfrac{4}{9} + 1\dfrac{3}{4} + 5\dfrac{8}{15} - 2\dfrac{1}{3} = $ _____

70. $6\dfrac{8}{15} + 2\dfrac{5}{9} + 1\dfrac{16}{21} - 3\dfrac{2}{5} = $ _____

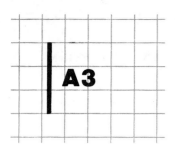

DECIMALS AND THEIR OPERATIONS

A3.1 PLACE VALUE AND ROUNDING

Numbers that are written to the *right* of the ones column in a place value chart are called **decimal numbers** and express *parts* of a **whole number**. A decimal point is placed between the ones column and the tenths column.

Example A

Ones	Decimal point	Tenths	Hundredths	Thousandths	Ten thousandths	Hundred thousandths
0	.	3	7	5	4	2

This number is read "thirty-seven thousand five hundred forty-two hundred thousandths" and can be written

$$(3 \times .1) + (7 \times .01) + (5 \times .001) + (4 \times .0001) + (2 \times .00001)$$

Example B

Ones	Decimal point	Tenths	Hundredths	Thousandths	Ten thousandths	
0	.	5				is read "five tenths"
0	.	6	3	8		is read "six hundred thirty-eight thousandths"
0	.	0	0	0	2	is read "two ten thousandths"
0	.	0	2	4		is read "twenty-four thousandths"
0	.	0	1			is read "one hundredth"

43

Rounding decimals: To round decimals, locate the digit in the position to be rounded.

Step 1. If the digit directly to its right is 5 or greater, increase the digit being rounded by 1 and drop all of the digits to its right.

Step 2. If the digit directly to its right is less than 5, leave the digit to be rounded unchanged and drop all of the digits to its right.

Example C   Round 6.0871 to the nearest hundredth.

⌐This 8 is to be rounded

6.0871 → 6.09

The 8 is to be rounded. The digit to its right is 5 or greater. Increase the 8 to 9. Drop the digits to its right.

Example D   Round 0.349 to the nearest tenth.

⌐This 3 is to be rounded

0.349 → 0.3

The 3 is to be rounded. The digit to its right is less than 5. Leave the 3 unchanged. Drop the digits to its right.

Example E   Round 4.95 to the nearest tenth.

⌐This 9 is to be rounded

4.95 → 5.0

The 9 is to be rounded. The digit to its right is 5. Increase the 9 by 1. Put 0 in the tenths place and add 1 to the 4. Drop the digit to the right of the tenths place.

A3.1 PROBLEMS

Write each number in the place value chart.

		Millions	Hundred thousands	Ten thousands	Thousands	Hundreds	Tens	Ones	Decimal point	Tenths	Hundredths	Thousandths	Ten thousandths	Hundred thousandths	Millionths
1.	Seven tenths								.	7					
2.	Seven thousandths								.	0	0	7			
3.	Seven thousand				7	0	0	0	.						
4.	Twelve hundredths								.	1	2				
5.	Eight and four tenths							8	.	4					
6.	One million seven hundred fifteen	1	0	0	0	7	1	5	.						
7.	Three hundred two millionths								.	0	0	0	2	0	2
8.	Twenty-five ten thousandths								.	0	0	2	5		
9.	Four hundred thousand and six hundredths		4	0	0	0	0	0	.	0	6				
10.	Five thousand one and twelve thousandths				5	0	0	1	.	0	1	2			

Give the place value of each 4.

11. 302.4 __Tenth__

12. 418.26 __Hundred__

13. 16.014 __Thousandth__

14. 4.56 __Ones__

15. 896.58734 __Hundred thousandth__

16. 1346.9 __Tens__

17. 4029.18 __Thousand__

18. 29.6427 __Hundredths__

19. 351.048 __Hundreth__

20. 2.673204 __Millionth__

45

Write as decimal numbers.

21. 2 thousandths _0.002_   22. 3 hundredths _0.03_

23. 15 hundredths _0.15_   24. 37 ten thousandths _0.0037_

25. 8 tenths _0.8_   26. 264 thousandths _0.264_

27. 500 and 6 tenths _500.6_   28. 2 and 7 thousandths _2.007_

29. 16 and 18 hundredths _16.18_

30. 54 hundred thousandths _0.054_

31. 3 tenths _0.3_

Round to the nearest tenth.

32. 0.638 _0.6_   33. 0.75 _.8_

34. 0.871 _0.9_   35. 0.9999 _1.0_

36. 15.927 _15.9_   37. 20.985 _21.0_

38. 106.808 _106.8_   39. 42.55 _42.6_

Round to the nearest hundredth.

40. 0.083_____   41. 0.1674_____   42. 0.9519_____

43. 6.0991_____   44. 4.355_____   45. 12.998_____

46. 0.6326_____   47. 0.5047_____

Round to the nearest thousandth.

48. 0.11115_____   49. 0.6372_____   50. 0.7787_____

51. 5.9399_____   52. 17.62591_____   53. 10.9999_____

54. 5.2525_____   55. 64.3752_____   56. 37.1001_____

57. 0.090909_____

A3.2 ADDITION AND SUBTRACTION

Addition: To add decimal numbers, line up the decimal points one under the other, in a straight line. Line up the digits, one under the other, according to their place value. Place a decimal point in the answer under the decimal points in the problem. Add as you would add whole numbers.

Example A   Find 3.1 + 6.58.

1. Line up decimal points.   3.1

2. Line up digits   +6.58
   in their place value   ‾‾‾‾‾
   columns.   9.68
3. Add as whole numbers.
4. Put decimal point in the
   answer.

Find 0.03 + 0.62 + 0.111.   Find 1.1 + 3.04 + 2.73.

   0.03   1.1
   0.62   3.04
+0.111   +2.73
‾‾‾‾‾‾   ‾‾‾‾‾
  0.761   6.87

Numbers are regrouped in decimal addition just as they arise in whole-number addition.

Example B   Find $0.83 + $0.68.

   1
$0.83   0.03 + 0.08 = 0.11
+$0.68
‾‾‾‾‾
$1.51   0.1 + 0.8 + 0.6 = 1.5

Find 0.02 + 0.68 + 0.41.   Find 3.2 + 0.75 + 0.074.

   1   1
  0.02   3.2
  0.68   0.75
+0.41   +0.074
‾‾‾‾‾   ‾‾‾‾‾‾
  1.11   4.024

Subtraction: To subtract decimal numbers, line up the decimal points and the digits according to their place value, placing the number being subtracted (minuend) *under* the number from which it is to be subtracted (subtrahend).

Example C   Find 0.962 − 0.84.

    1. Line up the decimal points.   0.962

    2. Line up digits in   − 0.84
        their place value columns.   0.122

    3. Subtract as you do
        whole numbers.

    4. Put decimal point
        in the answer.

Example D   Find 5.75 − 3.21.

$$\begin{array}{r} 5.75 \\ -3.21 \\ \hline 2.54 \end{array}$$

Example E   Find 1.64 − 0.02.

$$\begin{array}{r} 1.64 \\ -0.02 \\ \hline 1.62 \end{array}$$

Regrouping: Numbers are regrouped in decimal subtraction just as they are in whole-number subtraction. Place 0's to the right of the digits in the minuend when necessary so it will have as many decimal places as the subtrahend.

Example F   Find 17.6 − 12.84.

$$\begin{array}{r} 1\,7.6\,0 \leftarrow \text{Place a 0 here} \\ -1\,2.8\,4 \\ \hline 4.7\,6 \end{array}$$

Example G   Find \$5 − \$3.49.

$$\begin{array}{r} \$5.0\,0 \\ -\ 3.4\,9 \\ \hline \$1.5\,1 \end{array}$$

Example H   Find 2.653 − 1.7854.

$$\begin{array}{r} 2.6\,5\,3\,0 \\ -1.7\,8\,5\,4 \\ \hline 0.8\,6\,7\,6 \end{array}$$

A3.2 PROBLEMS

Add.

1. $0.1 + .01$

(handwritten work: 0.1 / 0.1 / 0.2)

2. $5.2 + 1.5$

3. $0.86 + 0.13$

(handwritten work: 1.725 / 4.060 / 3.100 / 8.885)

4. $3.1 + 4.06 + 1.725$

5. $0.23 + 67$

6. $8.19 + 6.42$

7. $\$3 + \$1.49 + \$6.07$

8. $0.56 + 0.9 + 2.83$

9. $2.8 + 7.1 + 3.4$

10. $9.728 + 0.05 + 1.627$

11. $\$0.03 + \$0.15 + \$0.26$

12. $0.005 + 0.5 + 0.05$

13. $3.4 + 0.237 + 4.04$

14. $0.54 + 0.165 + 0.26$

15. $37.31 + 9.091$

16.
$$\begin{array}{r} 45.2 \\ 111.11 \\ + 60.82 \\ \hline \end{array}$$

17.
$$\begin{array}{r} 89.063 \\ 1.8465 \\ + 22.301 \\ \hline \end{array}$$

18.
$$\begin{array}{r} 54.5 \\ 5.05 \\ + 16.67 \\ \hline \end{array}$$

19.
$$\begin{array}{r} 15.9632 \\ + 75.1508 \\ \hline \end{array}$$

20.
$$\begin{array}{r} \$203.03 \\ + 117.97 \\ \hline \end{array}$$

21.
$$\begin{array}{r} \$426.16 \\ + 138.02 \\ \hline \end{array}$$

22.
$$\begin{array}{r} 368.247 \\ 39.052 \\ + 104.106 \\ \hline \end{array}$$

23.
$$\begin{array}{r} 5.0634 \\ 2.1906 \\ + 7.8162 \\ \hline \end{array}$$

24.
$$\begin{array}{r} 18.09 \\ 24.101 \\ + 37.006 \\ \hline \end{array}$$

Subtract.

25. $0.47 - 0.25$ 26. $1.28 - 0.14$ 27. $3.5 - 1.2$

28. $11.8 - 7.24$ 29. $125.6 - 75.08$ 30. $99.09 - 45.90$

31. $\$115 - \$72.05$ 32. $\$50 - \$23.34$ 33. $\$12.19 - \$8.80$

34. $\$467.19$
    -238.63

35. $\$38.00$
    -29.75

36. $\$103.72$
    -78.93

37. 118.1
    -74.29

38. 212.731
    -98.624

39. 0.8354
    -0.2968

40. 0.624
    -0.26

41. 1.821
    -0.6437

42. $\$15.07$
    -9.68

43. 3.02 from 12.7   44. 6.81 from 7.5

45. 14.02 from 23.004   46. 0.86 from 1.09

47. 29.416 from 38   48. 19.005 from 23

A3.3 MULTIPLICATION AND DIVISION

Multiplication: To multiply decimals, multiply as if you were multiplying whole numbers. Then, determine the placement of the decimal point in the product.

To place the decimal point:

1. Add the number of decimal places in the numbers being multiplied. This sum tells you how many decimal places are in the product.
2. Moving from right to left, count off as many decimal places in the product as there are in the sum from Step 1, and place the decimal point accordingly.

Example A Find 3.7 × 2.   Find 1.36 × 0.5.

3.7	One decimal place		1.36	Two decimal places
× 2	No decimal places		× 0.5	One decimal place
7.4	One decimal place		0.680	Three decimal places

Example B Find 28.6 × 0.4.   Find 1.6 × 0.071.

28.6	One decimal place		1.6	One decimal place
× 0.4	One decimal place		×0.071	Three decimal places
11.44	Two decimal places		16	
			1120	
			0.1136	Four decimal places

After the decimal point has been placed in the product, 0's at the end of a decimal number may be dropped except when the product involves dollars and cents and you must have two places to the right of a decimal.

Example C $10.25
$$\times 0.8$$
$8.200 = $8.20

10.02		0.625
×5.05		×3.2
50.6010 = 50.601		2.0000 = 2

Sometimes there are not enough digits in the product to count off the required number of decimal places. When this occurs, you must use 0's as place holders.

Example D Find 0.2 × 0.2.   Find 0.01 × 0.3.

0.2	One decimal place		0.01	Two decimal places
×0.2	One decimal place		×0.3	One decimal place
0.04	Need two decimal places		0.003	Need three decimal places
↑	Use 0 as a place holder		↑	Use two 0's as place holders

Division: To divide a decimal by a whole number, place a decimal point in the quotient directly above the decimal point in the dividend. Then, divide as you would whole numbers.

Example E   Find $23.175 \div 15$.

First, place a decimal point in the quotient directly above the decimal point in the dividend.

$$15 \overline{)23.175}\,{}^{\textstyle .}$$

Then, divide as you would divide whole numbers.

$$
\begin{array}{r}
1.545 \\
15 \overline{)23.175} \\
-15 \\
\hline
81 \\
-75 \\
\hline
67 \\
-60 \\
\hline
75 \\
-75 \\
\hline
0
\end{array}
$$

To divide when a decimal point is in the divisor:

1. Change the divisor to a whole number by moving the decimal point to the right of the number.

2. Move the decimal point in the dividend to the right the same number of places as you moved the decimal point in the divisor. You may have to put 0's to the right of the last digit in the dividend to have a sufficient number of places. If there is no decimal point in the dividend, it is understood to be to the right of the last digit.

3. Place a decimal point in the quotient directly above the "new" decimal point in the dividend and divide as you would divide by a whole number.

Example F   Find $4.625 \div 0.25$.

First, move the decimal point in the divisor to the right. Then, move the decimal point in the dividend the same number of places and place a decimal point in the quotient directly above the "new" decimal in the dividend.

$$0.25 \overline{)4.62.5}$$

Then, divide as you would divide by a whole number.

$$
\begin{array}{r}
18.5 \\
25 \overline{)462.5} \\
-25 \\
\hline
212 \\
-200 \\
\hline
125 \\
-125 \\
\hline
0
\end{array}
$$

For every place in the dividend to the right of the decimal point, there must be a digit in the quotient. Use 0's as place holders when you cannot divide.

Not all division comes out even (that is, without a remainder). In the event this occurs, you may round your answer.

One way to round you answer is to carry the division one place farther than the place to which you desire the quotient to be rounded. Then use the rules for rounding decimals presented in 3.1.

Example G   Find $16.42 \div 5$ to the nearest tenth.

$$
\begin{array}{r}
3.28 = 3.3 \quad \text{(rounded)} \\
5\overline{)16.42} \\
-15 \\
\hline
14 \\
-10 \\
\hline
42 \\
-40 \\
\hline
2
\end{array}
$$

(8 is greater than 5. Add 1 to the 2 in the tenths place.)

Divide through the hundredths place; round to the nearest tenth.

Example H   Find $0.2461 \div 0.13$ to the nearest hundredth.

$$
\begin{array}{r}
1.893 = 1.89 \text{ (rounded)} \\
0.13\overline{)24.610} \rightarrow \text{place zero here to divide through the thousandths place} \\
-13 \\
\hline
116 \\
-104 \\
\hline
121 \\
-117 \\
\hline
40 \\
-30 \\
\hline
1
\end{array}
$$

(3 is less than 5. Leave the 9 in the hundredths place unchanged.)

Divide through the thousandths place; round to the nearest hundredth.

Another way to round an answer in division is to make use of the remainder. Divide through the place to which you desire your answer rounded, then look at the remainder. If the remainder is one-half or more than one-half of the divisor, add 1 to the digit in the last place you have divided. If the remainder is less than one-half of the divisor, leave the digits in the quotient unchanged.

Example I   Find $16.5 \div 12$ to the nearest tenth.

$$
\begin{array}{r}
1.3 = 1.4 \text{ (rounded)} \\
12\overline{)16.5} \\
-12 \\
\hline
45 \\
-36 \\
\hline
9
\end{array}
$$

The remainder, 9, is more than one-half of the divisor,

Add 1 to the 3 in the tenths place.

Example J   Find $6.847 \div 3.3$ to the nearest hundredth.

$$
\begin{array}{r}
2.07 \text{ (rounded)} \\
3.3\overline{)6.847} \\
-66 \\
\hline
247 \\
-231 \\
\hline
16
\end{array}
$$

The remainder, 16, is less than one-half of the divisor, 33.

Leave the quotient unchanged.

A3.3 PROBLEMS

Multiply.

1. 0.35×3 2. 263×0.2 3. 8.7×0.9

4. 1.63×0.7 5. 0.113×0.01 6. 2.5×0.25

7. 0.4×0.2 8. 0.5×0.5 9. 0.05×0.5

10. $\begin{array}{r} 4.2 \\ \times 1.7 \\ \hline \end{array}$ 11. $\begin{array}{r} 6.4 \\ \times 0.15 \\ \hline \end{array}$ 12. $\begin{array}{r} 0.81 \\ \times 0.09 \\ \hline \end{array}$

13. $\begin{array}{r} 10.4 \\ \times 0.27 \\ \hline \end{array}$ 14. $\begin{array}{r} 2.08 \\ \times 5.1 \\ \hline \end{array}$ 15. $\begin{array}{r} 0.613 \\ \times 0.85 \\ \hline \end{array}$

16. $\begin{array}{r} 0.417 \\ \times 0.38 \\ \hline \end{array}$ 17. $\begin{array}{r} 0.091 \\ \times 0.011 \\ \hline \end{array}$ 18. $\begin{array}{r} 0.078 \\ \times 0.022 \\ \hline \end{array}$

19. $\begin{array}{r} 0.1492 \\ \times 4.5 \\ \hline \end{array}$ 20. $\begin{array}{r} 30.64 \\ \times 0.34 \\ \hline \end{array}$ 21. $\begin{array}{r} 7.281 \\ \times 0.057 \\ \hline \end{array}$

Divide.

22. $26.35 \div 5$ 23. $5.67 \div 21$ 24. $360.4 \div 17$

25. $\$9.92 \div 8$ 26. $\$8.05 \div 35$ 27. $289.8 \div 46$

28. $5\overline{)0.455}$ 29. $7\overline{)0.588}$ 30. $12\overline{)7.308}$

31. $22\overline{)68.112}$ 32. $15\overline{)9.6}$ 33. $55\overline{)17.6}$

34. $1.68 \div 0.8$ 35. $0.245 \div 0.7$ 36. $0.341 \div 0.11$

37. $0.648 \div 0.12$ 38. $0.74 \div 0.005$ 39. $7.5 \div 0.06$

40. $1.5\overline{)2.25}$ 41. $0.24\overline{).1848}$ 42. $0.32\overline{)224}$

43. $0.55\overline{)275}$ 44. $0.12\overline{)24.36}$ 45. $0.38\overline{)3.914}$

46. $0.042\overline{)47.04}$ 47. $0.060\overline{)427.5}$ 48. $10\overline{)51.5}$

Divide. Round to the nearest hundredth.

49. $0.007\overline{)0.006575}$ 50. $0.014\overline{)0.46351}$ 51. $0.42\overline{)5.268}$

Divide. Round to the nearest thousandth.

52. $3\overline{)2.647}$ 53. $7\overline{)8.257}$ 54. $0.5\overline{)0.3826}$

Divide. Round to the nearest tenth.

55. $2.8\overline{)3.976}$ 56. $0.16\overline{)2.375}$ 57. $0.35\overline{)0.5082}$

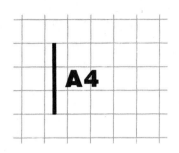

A4

EXPONENTIATION AND SQUARE ROOTS

A4.1 EXPONENTIATION

Exponents are used to tell how many times a number is to be used as a factor and are often used in solving problems.

Example A 3^2 means 3×3, or 9.

3 is used as a factor twice.

In this example, 3 is called the **base**,

2 is the **exponent**, and 3^2 is a **power**.

3^2 can be read as "three squared."

Example B 4^3 means $4 \times 4 \times 4$, or 64.

4 is used as a factor three times.

In this example, 4^3 is the third power

of 4, or the cube of 4, or "four cubed."

Example C 2^5 means $2 \times 2 \times 2 \times 2 \times 2$, or 32.

Example D $10 \times 10 \times 10$ can be written as 10^3.

Example E $7 \times 7 \times 7 \times 7$ can be written as 7^4.

Example F $3^2 \times 3^3 = (3 \times 3) \times (3 \times 3 \times 3) = 243$.

From this example, you can see that the answer could have been obtained by *adding* the exponents and then doing the multiplication.

$3^2 \times 3^3 = 3^5 = 243$ In general, $a^m \times a^n = a^{m+n}$.

This works only when the base numbers are the same.

Example G $\dfrac{2^5}{2^3} = \dfrac{\not2 \times \not2 \times \not2 \times 2 \times 2}{\not2 \times \not2 \times \not2} = 4$.

From this example, you can see that the answer could have been obtained by *subtracting* the exponents and then doing the multiplication.

$\dfrac{2^5}{2^3} = 2^{5-3} = 2^2 = 4$ In general, $\dfrac{a^m}{a^n} = a^{m-n}$ and $\dfrac{a^n}{a^n} = a^{n-n} = a^0 = 1$

for $a \neq 0$.

This works only when the base numbers are the same.

Example H $\left(\dfrac{2}{5}\right)^2 = \dfrac{4}{25}$.

Both the numerator and the denominator are to be squared in this case.

$\left(\dfrac{2}{5}\right)^2 = \dfrac{2^2}{5^2} = \dfrac{4}{25}$ In general, $\left(\dfrac{a}{b}\right)^n = \dfrac{a^n}{b^n}$.

Example I $(2^3)^4 = 2^3 \cdot 2^3 \cdot 2^3 \cdot 2^3 = 2^{12}$

$(2^3)^4 = 2^{3 \cdot 4} = 2^{12}$ In general, $(a^m)^n = a^{mn}$.

Example J $\left(\dfrac{2}{3}\right)^3 = \dfrac{8}{27}$.

Example K $\left(\dfrac{4}{5}\right)^2 + \left(\dfrac{3}{5}\right)^2 = \dfrac{16}{25} + \dfrac{9}{25} = \dfrac{25}{25} = 1$.

Example L $\left(\dfrac{3}{5}\right)^2 - \left(\dfrac{2}{5}\right)^2 = \dfrac{9}{25} - \dfrac{4}{25} = \dfrac{5}{25} = \dfrac{1}{5}$.

Example M $(0.2)^2 = (0.2)(0.2) = 0.04$.

Example N $(1.5)^2 = (1.5)(1.5) = 2.25$.

Example O $(0.3)^2 + (0.2)^2 = (0.3)(0.3) + (0.2)(0.2) = 0.09 + 0.04 = 0.13$.

A4.1 PROBLEMS

Compute.

1. $3^3 =$ ___27___
2. $4^2 =$ ___16___

3. $5^2 =$ ___25___
4. $7^3 =$ ___343___

5. $4^3 =$ ___64___
6. $3^4 =$ ___81___

7. $2^8 =$ ___256___
8. $8^2 =$ ___64___

9. $10^2 =$ ___100___
10. $10^4 =$ ___10,000___

11. $10^8 =$ ___100,000,000___
12. $10^5 =$ ___100,000___

13. $6^3 =$ ___216___
14. $9^4 =$ ___6561___

15. $7^5 =$ ___16,807___
16. $3^5 =$ ___243___

17. $\left(\dfrac{2}{3}\right)^3 =$ ___$\dfrac{8}{27}$___
18. $\left(\dfrac{4}{7}\right)^4 =$ ___$\dfrac{256}{2401} = .107$___

19. $\left(\dfrac{5}{8}\right)^2 =$ ___$\dfrac{25}{64}$___
20. $\left(\dfrac{9}{4}\right)^4 =$ ___$\dfrac{6561}{256} = 25.63$___

21. $(0.6)^2 =$ ___0.36___
22. $(0.3)^3 =$ ___0.027___

23. $(1.4)^2 =$ ___1.96___
24. $(2.5)^3 =$ ___15.625___

25. $\left(\dfrac{3}{8}\right)^2 + \left(\dfrac{1}{8}\right)^2 =$ ___$5/32$___
26. $\left(\dfrac{1}{6}\right)^3 + \left(\dfrac{5}{6}\right)^3 =$ ___0.483___

$\dfrac{9}{64} + \dfrac{1}{64} = \dfrac{10}{64} = \dfrac{5}{32}$
$\dfrac{1}{261} + \dfrac{125}{261} \quad \dfrac{126}{261}$

63

27. $\left(\dfrac{4}{7}\right)^2 - \left(\dfrac{2}{7}\right)^2 =$ ___12/49___

$\dfrac{16}{49} - \dfrac{4}{49} = \dfrac{12}{49}$

28. $\left(\dfrac{5}{9}\right)^3 - \left(\dfrac{2}{9}\right)^3 =$ _____

29. $(2^3)^5 =$ ___$2^{15} = 32,768$___

30. $(5^2)^3 =$ ___5^6___

31. $3^5 \times 3^4 =$ ___3^9___

32. $2^4 \times 2^5 =$ ___2^9___

33. $10^5 \times 10^3 =$ ___10^8___

34. $4^3 \times 4^5 =$ ___4^8___

35. $\dfrac{12^9}{12^5} =$ ___12^4___

36. $\dfrac{3^4}{3^2} =$ _____

37. $\dfrac{6^{24}}{6^{19}} =$ ___6^5___

38. $\dfrac{10^{10}}{10^8} =$ _____

39. $(2 + 1)^3 =$ ___27___

40. $(5 + 4)^2 =$ ___9 81___

41. $(8 - 7)^5 =$ ___$1^5 = 1$___

42. $(9 - 4)^3 =$ ___5 125___

43. $(8 + 7)^2 =$ ___$15^2 = 225$___

44. $(9 - 3)^3 =$ ___6 216___

45. $(8)^2 + (7)^2 =$ ___113___
$64 + 49$

46. $(9)^3 - (3)^3 =$ _____

47. $(9)^3 + (4)^3 =$ ___793___

48. $(5)^2 + (4)^2 =$ ___41___
$25 + 16$

49. $(3 + 2)^2 + (4 - 1)^3 =$ ___52___
$25 + 27$

50. $(7 - 4)^4 + (2 - 1)^3 =$ ___82___
$3^4 + 1$

51. $\left(\dfrac{2}{3}\right)^3 + \left(\dfrac{4}{9}\right)^2 =$ ___40/81___

$\dfrac{8}{27} + \dfrac{16}{81}$ $\dfrac{24}{81} + \dfrac{16}{81} = \dfrac{40}{81}$

52. $\left(\dfrac{3}{5}\right)^4 - \left(\dfrac{2}{5}\right)^3 =$ _____

A4.2 SQUARE ROOTS

Squaring a number means multiplying that number by itself. The process can be visualized by thinking about the area of a square surface. The symbol for squaring a number is the exponent 2. Any number with the exponent 2 may be thought of as the square of that number.

Example A $3^2 = 9.$

As a pictorial representation, this might be:

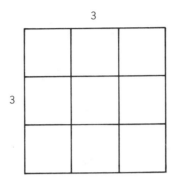

Example B $5^2 = 25.$

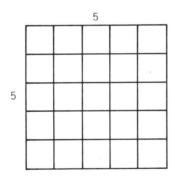

To find the **square root** of a number, we look for a number that when multiplied by itself will equal the given number. The symbol for square root is "$\sqrt{}$". This symbol is called a **radical sign**.

Example C The square root of 25 is 5 because $5 \times 5 = 25.$ $\sqrt{25} = 5$

Example D $\sqrt{81} = 9$ because $9 \times 9 = 81.$

Example E $\sqrt{0.25} = 0.5$ because $0.5 \times 0.5 = 0.25.$

Example F $\sqrt{0.04} = 0.2$ because $0.2 \times 0.2 = 0.04.$

65

Example G $\sqrt{\dfrac{9}{16}} = \dfrac{3}{4}$ because $\dfrac{3}{4} \times \dfrac{3}{4} = \dfrac{9}{16}$.

Below is a partial table of whole numbers and their squares.

Number	0	1	2	3	4	5	6	7	8	9	10	11	12	13	14	15
Its square	0	1	4	9	16	25	36	49	64	81	100	121	144	169	196	225

This table can also be used to find the square roots of the numbers in the bottom row.

Example H   Find the square root of 196.

Look for 196 in the bottom row. Then look for the number directly above it. 14 is the number above 196. $\sqrt{196} = 14$.

The numbers in the bottom row are called **perfect squares**. They have whole number square roots. Only perfect squares have whole number square roots.

Finding the square root of a whole number that is not a perfect square requires either a calculator, the use of a computational process, or a good sense of estimation. The square root of a whole number that is not a perfect square can be approximated. Let us use the chart to find between which two whole numbers the square root of another number is located.

Example I   Between which two whole numbers is the square root of 33? Looking at the bottom row, you can see that the number 33 is between 25 and 36. Since the square root of 25 is 5 and the square root of 36 is 6, the square root of 33 is between 5 and 6. In fact, $\sqrt{33} \approx 5.745$.

Example J   Between which two whole numbers is the square root of 72?

In the table of squares, 72 is between 64 and 81.

$\sqrt{64} = 8$, $\sqrt{81} = 9$. $\sqrt{72}$ is between 8 and 9.

Example K $\sqrt{250}$ is between which two whole numbers?

250 is between 225 and 256; $15^2 = 225$, $16^2 = 256$. So $\sqrt{250}$ is between 15 and 16.

A4.2 PROBLEMS

Find the indicated square roots.

1. $\sqrt{81}$ = ___9___   2. $\sqrt{25}$ = ___5___

3. $\sqrt{36}$ = ___6___   4. $\sqrt{49}$ = ___7___

5. $\sqrt{64}$ = ___8___   6. $\sqrt{121}$ = ___11___

7. $\sqrt{169}$ = ___13___   8. $\sqrt{225}$ = ___15___

9. $\sqrt{324}$ = ___18___   10. $\sqrt{361}$ = ___19___

11. $\sqrt{144}$ = ___12___   12. $\sqrt{484}$ = ___22___

13. $\sqrt{729}$ = ___27___   14. $\sqrt{256}$ = ___16___

15. $\sqrt{10,000}$ = ___100___   16. $\sqrt{1}$ = ___1___

17. $\sqrt{0.09}$ = ___0.3___   18. $\sqrt{0.16}$ = ___0.4___

19. $\sqrt{0.0004}$ = ___0.02___ 20. $\sqrt{0.0009}$ = ___0.03___

21. $\sqrt{\dfrac{25}{36}}$ = ___$\frac{5}{6}$___   22. $\sqrt{\dfrac{9}{49}}$ = ___$\frac{3}{7}$___

23. $\sqrt{\dfrac{4}{81}}$ = ___$\frac{2}{9}$___   24. $\sqrt{\dfrac{16}{25}}$ = ___$\frac{4}{5}$___

25. $\sqrt{\dfrac{144}{81}}$ = ___$\frac{12}{9} = 3$___   26. $\sqrt{\dfrac{100}{25}}$ = ___2___

Tell between what two whole numbers the square root of the given number lies.

27. 7 _2_ and _3_   28. 13 _3_ and _4_

29. 24 _4_ and _5_   30. 59 _7_ and _8_

31. 72 _8_ and _9_   32. 88 _9_ and _10_

33. 135 _11_ and _12_   34. 250 _15_ and _16_

35. 604 _24_ and _25_   36. 621 _24_ and _25_

37. 528 _22_ and _23_   38. 401 _20_ and _21_

39. 8 _2_ and _3_   40. 80 _8_ and _9_

41. 800 _28_ and _29_   42. 8000 _89_ and _90_

43. 5 _2_ and _3_   44. 50 _7_ and _8_

45. 500 _22_ and _23_   46. 5000 _70_ and _71_

47. 10 _3_ and _4_   48. 1000 _31_ and _32_

49. 100,000 _316_ and _317_   50. 10,000,000 _3162_ and _3163_

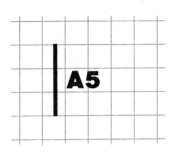

A5 FRACTION-DECIMAL CONVERSION

A5.1 FRACTIONS TO DECIMALS

To change a fraction to a decimal, divide the numerator by the denominator.

Example A Change $\frac{3}{5}$ to a decimal.

$\frac{3}{5}$ means $3 \div 5$. To divide 3 by 5, put a decimal point after the 3 with as many 0's as are needed. The decimal point in the answer will be directly above the decimal point placed after the 3.

$$\begin{array}{r} 0.6 \\ 5\overline{\smash{)}3.0} \\ \underline{30} \\ 0 \end{array} \qquad \frac{3}{5} = 0.6$$

Example B Change $\frac{7}{20}$ to a decimal.

$$\begin{array}{r} 0.35 \\ 20\overline{\smash{)}7.00} \\ \underline{60} \\ 100 \\ \underline{100} \\ 0 \end{array} \qquad \frac{7}{20} = 0.35$$

Example C Change $\frac{5}{8}$ to a decimal.

$$\begin{array}{r} 0.625 \\ 8\overline{\smash{)}5.000} \\ \underline{48} \\ 20 \\ \underline{16} \\ 40 \\ \underline{40} \\ 0 \end{array} \qquad \frac{5}{8} = 0.625$$

Example D   Change $\frac{5}{4}$ to a decimal.

$$\begin{array}{r} 1.25 \\ 4\overline{)5.00} \\ \underline{4} \\ 10 \\ \underline{8} \\ 20 \\ \underline{20} \\ 0 \end{array} \qquad \frac{5}{4} = 1.25$$

The mixed number $1\frac{1}{4}$ could have been used instead of $\frac{5}{4}$.

Change only the $\frac{1}{4}$ to a decimal. $\begin{array}{r} 0.25 \\ 4\overline{)1.00} \end{array}$

Add the 1 to 0.25 $1 + 0.25 = 1.25$.

In Examples A–D, the remainders are all 0. This does not always happen.

Example E   Change $\frac{27}{22}$ to a decimal.

$$\begin{array}{r} 1.22727 \\ 22\overline{)27.00} \\ \underline{22} \\ 50 \\ \underline{44} \\ 60 \\ \underline{44} \\ 160 \\ \underline{154} \\ 60 \\ \underline{44} \\ 160 \\ \underline{154} \\ 6 \end{array}$$

The remainder is again 6, and you can see that this decimal will continue forever, with the digits 27 repeated endlessly, and is written $1.2\overline{27}$. The bar indicates the decimals that are repeated. We use rounding to give any desired degree of accuracy. To the nearest tenth, the answer is 1.2, to the nearest hundredth, it is 1.23, to the nearest thousandth, it is 1.227, and so on.

To change a fraction in which the denominator is a power of ten (10, 100, 1000, etc.) to a decimal, just place a decimal point in the numerator which gives it the same power of ten that is in the denominator.

Both $\dfrac{7}{100}$ and 0.07 are read "seven hundredths."

Example F   Change $\dfrac{7}{100}$ to a decimal.

$$\frac{7}{100} = 0.07$$

Example G   Change $\dfrac{9}{1000}$ to a decimal.

$$\frac{9}{1000} = 0.009$$

Example H   Add: $5 + \dfrac{3}{10} + \dfrac{7}{1000}$.

$$5 + 0.3 + 0.007 = 5.307$$

Example I   Add: $\dfrac{5}{4} + 0.35$.

$$1.25 + 0.35 = 1.60 = 1.6$$

Example J   Subtract: $\dfrac{5}{4} - \dfrac{7}{20}$.

This can be done in two ways. In terms of fractions,

$$\frac{5}{4} - \frac{7}{20} = \frac{25}{20} - \frac{7}{20}$$

$$= \frac{18}{20} = \frac{9}{10}$$

In terms of decimals,

$$\frac{5}{4} - \frac{7}{20} = 1.25 - 0.35$$

$$= 0.9$$

Example K   Use decimals to compute

$\dfrac{7}{8} - 0.625$.

$$\frac{7}{8} - 0.625 = 0.875 - 0.625$$

$$= 0.25$$

There is sometimes another way to convert fractions to decimals.

Example L   Change $\frac{7}{20}$ to a decimal.

This is Example B again, but now we do it without dividing 7 by 20. Multiply numerator and denominator by 5. This is just multiplying by 1.

$$\frac{7}{20} = \frac{7 \times 5}{20 \times 5}$$

$$= \frac{35}{100} = 0.35$$

Example M   Change $\frac{5}{4}$ to a decimal.

This is Example D again, but now we multiply numerator and denominator by 25.

$$\frac{5}{4} = \frac{5 \times 25}{4 \times 25}$$

$$= \frac{125}{100} = 1.25$$

Example N   Change $\frac{17}{125}$ to a decimal.

You can divide, as before. But you can also realize that 125 is the product of three 5s. Thus, if the numerator and denominator are multiplied by the product of three 2s, then the denominator will be the product of three 10s.

$$\frac{17}{125} = \frac{17}{5^3} \times \frac{2^3}{2^3}$$

$$= \frac{17 \times 8}{10^3}$$

$$= \frac{136}{1000} = 0.136$$

A5.1 PROBLEMS

Write each fraction and mixed number as a decimal.

1. $\dfrac{7}{2} =$ _____

2. $\dfrac{17}{5} =$ _____

3. $\dfrac{93}{25} =$ _____

4. $\dfrac{27}{4} =$ _____

5. $\dfrac{67}{50} =$ _____

6. $\dfrac{31}{20} =$ _____

7. $\dfrac{6}{5} =$ _____

8. $\dfrac{5}{2} =$ _____

9. $\dfrac{13}{4} =$ _____

10. $\dfrac{71}{25} =$ _____

11. $\dfrac{11}{20} =$ _____

12. $\dfrac{33}{50} =$ _____

13. $\dfrac{9}{2} =$ _____

14. $\dfrac{13}{5} =$ _____

15. $\dfrac{34}{25} =$ _____

16. $\dfrac{9}{4} =$ _____

17. $\dfrac{27}{12} =$ _____

18. $\dfrac{102}{75} =$ _____

19. $\dfrac{3}{8} =$ _____

20. $\dfrac{29}{125} =$ _____

21. $\dfrac{3}{250} =$ _____

22. $\dfrac{7}{40} =$ _____

23. $\dfrac{49}{280} =$ _____

24. $\dfrac{9}{750} =$ _____

25. $3\dfrac{5}{8} =$ _____

26. $5\dfrac{3}{8} =$ _____

27. $1\dfrac{3}{4} =$ _____

28. $3\dfrac{1}{4} =$ _____

29. $11\dfrac{5}{16} =$ _____

30. $5\dfrac{11}{16} =$ _____

31. $9\dfrac{7}{20} =$ _____

32. $7\dfrac{9}{20} =$ _____

33. $8\dfrac{11}{20} =$ _____

34. $\dfrac{201}{75} =$ _____

35. $\dfrac{33}{44} =$ _____

36. $\dfrac{41}{125} =$ _____

37. $\dfrac{5}{8} =$ _____

38. $\dfrac{3}{40} =$ _____

39. $\dfrac{7}{250} =$ _____

40. $\dfrac{75}{24} =$ _____

41. $\dfrac{24}{75} =$ _____

42. $\dfrac{2}{625} =$ _____

43. $\dfrac{3}{16} =$ _____

44. $\dfrac{5}{32} =$ _____

45. $\dfrac{3}{3125} =$ _____

46. $\dfrac{12}{750} =$ _____

47. $\dfrac{42}{280} =$ _____

48. $\dfrac{3}{9375} =$ _____

49. $\dfrac{28}{224} =$ _____

50. $\dfrac{91}{140} =$ _____

51. $\dfrac{222}{450} =$ _____

52. $\dfrac{3}{10} =$ _____

53. $\dfrac{17}{100} =$ _____

54. $\dfrac{15}{1000} =$ _____

55. $\dfrac{118}{100} =$ _____

56. $\dfrac{17}{10} =$ _____

57. $\dfrac{136}{100} =$ _____

58. $5\dfrac{7}{1000} =$ _____

59. $14\dfrac{12}{100} =$ _____

60. $8\dfrac{9}{10} =$ _____

Find the decimal answer.

61. $1 + \dfrac{3}{10} + \dfrac{7}{100} =$ _____

62. $2 + \dfrac{9}{100} + \dfrac{9}{1000} =$ _____

63. $\dfrac{6}{10} + \dfrac{8}{100} + \dfrac{10}{1000} =$ _____

64. $\dfrac{3}{10} + \dfrac{7}{100} + \dfrac{9}{1000} =$ _____

65. $\dfrac{8}{7} + 6.66 =$ _____

66. $\dfrac{9}{4} + 1.33 =$ _____

67. $\dfrac{5}{8} + 0.125 =$ _____

68. $\dfrac{3}{8} + 1.125 =$ _____

69. $\dfrac{3}{4} + 0.75 =$ _____

70. $\dfrac{12}{5} + 2.5 =$ _____

A5.2 DECIMALS TO FRACTIONS

To change a decimal to a fraction, write the digits in the fraction without the decimal point as the numerator and write the power of ten that expresses the place value of the decimal as the denominator.

Example A   Change 0.11 to a fraction.

$$0.11 = \frac{11}{100}$$

Both the fraction *and* the decimal are read "eleven hundredths."

Example B   Change 0.008 to a fraction.

$$0.008 = \frac{8}{1000}$$

Both the fraction and the decimal are read "eight thousandths." $\frac{8}{1000}$ can be rewritten in lower terms.

$$\frac{8}{1000} = \frac{\cancel{2} \cdot \cancel{2} \cdot \cancel{2}}{\cancel{2} \cdot \cancel{2} \cdot \cancel{2} \cdot 5 \cdot 5 \cdot 5} = \frac{1}{5^3} = \frac{1}{125}$$

So, $0.008 = \frac{8}{1000} = \frac{1}{125}$.

Example C   Change 3.3 to a fraction.

$$3.3 = \frac{33}{10}$$

$\frac{33}{10}$ can be rewritten as a mixed number.

$$\frac{33}{10} = 3\frac{3}{10}$$

So, $3.3 = 3\frac{3}{10}$. The decimal and fraction are both read "three and three tenths."

Example D   Change 7.005 to a fraction.

$$7.005 = \frac{7005}{1000} \qquad \frac{7005}{1000} \text{ can be rewritten as a mixed number.}$$

$$\frac{7005}{1000} = 7\frac{5}{1000}$$

$\frac{5}{1000}$ can be rewritten in lower terms.

$$\frac{5}{1000} = \frac{5 \times 1}{5 \times 200} = \frac{1}{200}$$

So, $7.005 = \frac{7005}{1000} = 7\frac{5}{1000} = 7\frac{1}{200}$.

Example E   Change 11.45 to a mixed number.

$$11.45 = \frac{1145}{100} = \frac{5 \times 229}{5 \times 20} = \frac{229}{20} = 11\frac{9}{20}$$

Decimal numbers greater than 1 can be written as mixed numbers by leaving the whole number part alone and working with only the fractional part.

$$11.45 = 11\frac{45}{100} = 11\frac{9}{20}$$

Sometimes, the numerator and/or denominator of the fraction may themselves be fractions or decimals.

Example F   Write $\dfrac{\left(\dfrac{9}{20}\right)}{0.5}$ as a fraction and as a decimal.

Change either the $\dfrac{9}{20}$ to a decimal or the 0.5 to a fraction.

We shall solve it both ways.

1. Change $\dfrac{9}{20}$ to a decimal.

$$\begin{array}{r} 0.45 \\ 20\overline{)9.00} \\ 80 \\ \hline 100 \\ 100 \\ \hline 0 \end{array}$$

Then the problem becomes $\dfrac{0.4\cancel{5}^{9}}{0.\cancel{5}_{1}} = 0.9 = \dfrac{9}{10}$

2. Change 0.5 to a fraction.

$$0.5 = \frac{5}{10} = \frac{1}{2}$$

Then the problem becomes

$$\frac{\dfrac{9}{20}}{\dfrac{1}{2}} = \frac{9}{20} \div \frac{1}{2} = \frac{9}{20} \times \frac{2}{1} = \frac{18}{20} = \frac{9}{10} = 0.9$$

The result is $\dfrac{\left(\dfrac{9}{20}\right)}{0.5} = \dfrac{9}{10} = 0.9.$

A5.2 PROBLEMS

(handwritten notes in left margin:)

1. Write digits in decimal in the Numer. w/o dec. pt.
2. Den. will be a 1 w/same # of zeros as places to the right of dec. pt. in #1

Example:
.02948
$\dfrac{2948}{100000}$
$= \dfrac{737}{25,000}$

$.8 = \dfrac{8}{10} = \dfrac{4}{5}$

Write fraction + decimal answer

$\left[\dfrac{\left(\dfrac{5}{8}\right)}{.5}\right]$ $\dfrac{\dfrac{5}{8}}{\dfrac{5}{10}} = \dfrac{5}{8} \cdot \dfrac{10}{5}$ $\boxed{\dfrac{5}{2}}$

$\dfrac{0.15}{100}$ 3

Write each as a fraction. Reduce when possible.

1. 0.2 = _____ $\dfrac{1}{5}$ _____   2. 0.5 = _____

3. 0.15 = _____ 3/20 _____   4. 0.16 = _____

5. 1.28 = _____   6. 1.25 = _____

7. 1.005 = _____   8. 1.008 = _____

9. 2.56 = _____   10. 6.25 = _____

11. 0.3125 = _____   12. 0.1024 = _____

13. 0.375 = _____   14. 0.625 = _____

15. 0.875 = _____   16. 0.125 = _____

17. 1.12 = _____   18. 1.28 = _____

19. 0.02948 = _____   20. 0.15625 = _____

21. 2.4024 = _____   22. 65.1065 = _____

23. 85.3585 = _____   24. 48.1684 = _____

25. 0.0036 = _____   26. 0.0092 = _____

27. 0.109375 = _____   28. 0.46875 = _____

29. 0.14375 = _____   30. 0.11875 = _____

31. 12.2125 = _____   32. 24.4625 = _____

33. 123.456 = _____   34. 345.679 = _____

35. 876.543 = _____   36. 654.321 = _____

37. $0.007168 =$ _____

38. $0.08432 =$ _____

39. $0.0390625 =$ _____

40. $0.0078125 =$ _____

41. $0.000032 =$ _____

42. $0.000025 =$ _____

43. $25.6000128 =$ _____

44. $51.2000256 =$ _____

Write each decimal number as a mixed number.

45. $1.25 =$ _____

46. $1.28 =$ _____

47. $1.008 =$ _____

48. $1.005 =$ _____

49. $6.25 =$ _____

50. $2.56 =$ _____

51. $1.28 =$ _____

52. $1.12 =$ _____

53. $65.105 =$ _____

54. $2.4025 =$ _____

55. $122.1924 =$ _____

56. $437.3125 =$ _____

57. $15.4515 =$ _____

58. $64.3264 =$ _____

59. $48.1684 =$ _____

60. $85.3585 =$ _____

Write each answer as both a fraction and as a decimal.

61. $\dfrac{\left(\frac{3}{20}\right)}{0.6} =$ _____

62. $\dfrac{\left(\frac{7}{20}\right)}{0.07} =$ _____

63. $\dfrac{\left(\frac{5}{8}\right)}{0.5} =$ _____

64. $\dfrac{\left(\frac{7}{8}\right)}{0.08} =$ _____

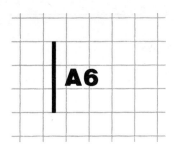

A6

APPLICATIONS

A6.1 AVERAGES

Often, useful information is expressed as an **average**. Finding an average (also called the arithmetic mean) involves both a sum and a quotient.

To find an average (mean):

1. Add all of the numbers in the group;

2. Divide that sum by the number of items in the group.

Example A   Jose's bowling scores today were 165, 150, and 162. What was his average score?

1. Add the scores. 165 + 150 + 162 = 477.

2. Divide the sum of the scores (477) by the number of scores (3).

   $477 \div 3 = 159$

Jose's bowling average was 159.

Example B   A basketball player scored the following number of points per game.

First game—23   Fifth game—16
Second game—19   Sixth game—21
Third game—17   Seventh game—23
Fourth game—25   Eighth game—24

What was his scoring average per game for the season?

23 + 19 + 17 + 25 + 16 + 21 + 23 + 24 = 168

Number of games → 8$\overline{)168}$ ← Total number of points, quotient 21

His average was 21 points per game.

Example C   Jeffrey's test scores on ELM practice tests are: 75, 90, 83, 92. What is his average test score?

$$\frac{75 + 90 + 83 + 92}{4} = 85 \text{ average test score}$$

Example D   During one week, the high temperatures for each day expressed in degrees Celsius recorded in an Alaskan city were $-5°$, $17°$, $-3°$, $0°$, $-9°$, $12°$, $2°$. What was the average high temperature during that week?

$$\frac{(-5) + (17) + (-3) + (0) + (-9) + (12) + (2)}{7} = \frac{14}{7} = 2$$

The average high temperature for the week was 2°C.

Example E   Lauren drove 153 miles in 3 hours. What was her average speed in miles per hour?

In this problem, the number of miles Lauren drove during each of the 3 hours has already been determined. You need only to do the division to find her average speed per hour.

$$153 \div 3 = 51$$

Lauren's average speed was 51 miles per hour.

Example F   The County Junior College basketball team scored a total of 836 points during its 10-game season. What was the average number of points scored per game?

$$\frac{836}{10} = 83.6 \text{ average number of points scored per game}$$

A6.1 PROBLEMS

1. The following high temperatures in Celsius were recorded each day for one week:
37°, 36°, 32°, 26°, 25°, 33°, 28°. What was the average high temperature for that week? _____

2. Forty-seven students handed out 1081 circulars advertising the school play. What is the average number of circulars handed out by each student? _23 circulars_

   1. Average number
   2. $\frac{1081}{47}$ = Avg. number   23
   3. Solve it
   4. check

3. Assuming the sales were the same each month, how many dictionaries did the bookstore sell in one month if they sold 2928 in one year? _____

4. The starting players on the high school football team weighed the following numbers of pounds: 212, 207, 203, 197, 200, 224, 190, 201, 199, 213, 198. Find the average weight of the starting lineup. _204_

   1. Add up weights
   2. Divide weights by 11
   3. check

5. Average these test scores: 79, 82, 70, 91, 65, 73, 86. ___78___

6. For the first six games of the soccer season, a team scored the following number of goals: 5, 3, 1, 0, 4, 2. What is their average number of goals scored per game? _____

7. Stephan drove 242 kilometers in 4 hours. What was his average driving speed in kilometers per hour? _____

8. A city's low Fahrenheit temperatures for one week are shown below. Find the average low temperature for the week. ___26°F___

S	M	T	W	T	F	S
27°	31°	29°	20°	19°	23°	33°

1. Add Su thru Saturday $\frac{182}{7}$
2. Divide by 7
3. Check

9. Kim made the following scores on mathematics tests during the semester: 83, 97, 62, 72, 93, 68, 71. What was her average test score? _____

10. Sue read the following number of library books each month of last year. What was the average number of books she read each month? ___11.5___

Jan.	Feb.	Mar.	Apr.	May	June	July	Aug.	Sept.	Oct.	Nov.	Dec.
15	18	12	0	19	7	20	14	13	8	7	5

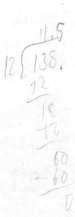

A6.2 PERCENTS

Percent is a way to express a quantity. Percent is symbolized by "%", which means "hundredth." To write a percent as a fraction, divide by 100 and remove the % sign.

Example A   Express 17% as a fraction.   Express 4% as a fraction.

$$17\% = \frac{17}{100} \qquad\qquad 4\% = \frac{4}{100} = \frac{1}{25}$$

Example B   Express 117% as a fraction.

$$117\% = \frac{117}{100} = 1\frac{17}{100}$$

Percents can also be expressed as decimals. To change a percent to a decimal, move the decimal point two places to the left, using 0's if necessary, and drop the percent sign. If there is no written decimal point, it is understood to be to the right of the last digit.

Example C   Express 18% as a decimal.   Express 2% as a decimal.

$$18\% = 0.18 \qquad\qquad 2\% = 0.02$$

Example D   Express 13.6% as a decimal.

$$13.6\% = 0.136$$

Decimals and fractions can be written as percents by performing the reverse operations.

Example E $\dfrac{1}{2} = \dfrac{50}{100} = 0.50 = 50\%$

Example F   Express the given decimals as percents.

Decimal	Percent
0.02	2%
0.25	25%
1.3	130%

Example G   Express the given fractions as percents.

Fraction	Percent
$\dfrac{8}{100}$	8%
$\dfrac{108}{100}$	108%
$\dfrac{3}{10}$	30%

83

Example H Express as equivalents.

Fraction	Decimal	Percent
$\frac{3}{4}$	0.75	75%
$\frac{1}{200}$	0.005	0.5%
$\frac{11}{8}$	1.375	137.5%

To find a specified percent of a number, change the percent to either a fraction or a decimal and multiply it by the number.

Example I Find 15% of 87.

1. Change 15% to a decimal.

$15\% = 0.15$

2. Multiply 0.15 by 87.

$0.15 \times 87 = 13.05$

15% of $87 = 13.05.$

Example J Twenty-four percent of the students in a mathematics class are female. There are 75 students in the class. How many of them are female?

$24\% = 0.24$

$0.24 \times 75 = 18$

18 of the students in the class are female.

Sometimes, it is necessary to find what percent one number is of another number.

Example K 18 is what percent of 20?

1. Divide 18 by 20.

$18 \div 20 = 0.9$

2. Change 0.9 to a percent.

$0.9 = 90\%$

18 is 90% of 20.

Example L 75 is what percent of 50?

1. Divide 75 by 50.

$75 \div 50 = 1.5$

2. Change 1.5 to a percent.

$1.5 = 150\%$

75 is 150% of 50.

A6.2 PROBLEMS

Handwritten notes (left margin):

Percent → Decimal
1. move decimal 2 pts to the left

Example:
1. .80% = .8
2. 17% = .17
3. .3% = .03
4. 185% = 1.85

3 types of percent problems
1) A is what % of b
 - Solve by dividing a by b
 - Change to a percent.
2) Find A % of B.
 - change % to decimal. multiply times B.

Find 15% of 75

(.15 × 75)
[11.25]

3) A is b% of what number
 - Change % to decimal mult
8 is 25% of what number

Write each number in the form indicated.

1. 3.45 = __345 %__ (percent)   2. 5.52 = _____ (percent)

     345 : 69 / 10000 20,000

3. 3.45% = _____ (fraction)   4. 5.52% = _____ (decimal)

5. 3.45% = _____ (decimal)   6. 5.52% = _____ (fraction)

7. 0.25 = _____ (percent)   8. 0.75 = _____ (percent)

9. 0.25 = _____ (fraction)   10. 0.75% = _____ (decimal)

11. 0.25% = _____ (decimal)   12. 0.75% = _____ (fraction)

13. $\frac{86}{100}$ = _____ (percent)   14. $\frac{1}{100}$ = _____ (percent)

15. $\frac{5}{10}$ = _____ (percent)   16. $\frac{7}{10}$ = _____ (percent)

17. $\frac{3}{4}$ = _____ (percent)   18. $\frac{3}{8}$ = _____ (percent)

19. $\frac{1}{50}$ = _____ (percent)   20. $\frac{3}{50}$ = _____ (percent)

Answer.

21. The enrollment in English 201 is 515 students. Of these, 80% are making a passing grade. How many are passing? _____

22. The enrollment in Mathematics 102 is 430 students. Of these, 70% are making C or better. How many are making C or better? _____

85

23. In the Scuba Divers Club, there are 65 members, and 13 are making a dive trip to Mazatlán. What percent of the club is going? _____

24. Of the 85 members in the Skydiver's Club, 34 will make a jump this weekend. What percent will jump this weekend? _____

25. The sales tax in Silvertown is 5%. How much sales tax is there on an item that costs $210? _____

26. The VAT tax in London is 6%. How much VAT tax is there on an item that costs $170? $10.20 _____

$$170(.06) = \$10.20$$

27. The precipitation last year was 120% of the normal precipitation of 35 inches. What was the precipitation last year? _____

28. 16 is what percent of 20? _____

29. 20 is what percent of 16? _____

30. Bob's house is worth 225% of its original cost of $76,240. How much is his house worth now? _____

31. On a 110-question test, students must answer correctly at least 66 questions to pass. In order to earn a passing score, what is the minimum percent of the questions that must be answered correctly? _____

32. On a 75-question test, students must answer correctly at least 45 questions to pass. In order to earn a passing score, what is the minimum percent of the questions that must be answered correctly? _____

33. In the Ultimate Frisbee Club, 55% of the members are seniors, and the rest are sophomores and juniors. In all, there are 18 members who are either sophomores or juniors. How many people are in the club? _____

18 is 45% of what number?

34. A $37.95 item is on sale at a discount of 20%. The sales tax is 8% of the sales price. What will be the total that one must pay for the item? $32.79

1.) Find the discount (.2)(37.95) $7.59

2. Find sales price - 37.95 - discount
   $$\frac{7.59}{\$30.36}$$

3. Find sales Tax 30.36(.08) = 2.4288 = $2.43

A6.3 OTHER APPLICATIONS

Many mathematics problems are what we call **word problems**. Word problems are real-life situations that you turn into mathematical problems. You find the solutions, and then interpret the answers in terms of the original settings.

Before you begin solving a word problem, there are some steps you should take.

1. Read the problem carefully.

2. Note the facts and numbers with which you will be working.

3. Understand the question being asked.

4. Look for key words that will help you decide which operation or operations you will use to solve the problem.

After solving the problem, check to see if the answer "fits"—that it is a reasonable solution. A computational error or error in placing a decimal point will cause you to get an incorrect answer. It is possible that there are a variety of ways to solve a problem, all of which lead to the correct solution.

Example A If 3 pounds of beans cost $2.16, what is the cost of 5 pounds of beans?

One way to solve this problem is to find the cost of 1 pound of beans by dividing $2.16 by 3, and then multiplying this answer by 5.

$2.16 ÷ 3 = $0.72

$0.72 × 5 = $3.60

Five pounds of beans would cost $3.60.

Example B A turkey is to be cooked 30 minutes for each pound of weight. If a turkey weighing 12 pounds is to be ready for dinner at 7:00 P.M., what time should it be put in the oven?

You could think: If it takes 30 minutes per pound to cook the turkey, then it takes 1 hour per 2 pounds. The weight of the turkey (12 pounds) divided by 2 pounds per hour is 6 hours.

Now, subtract 6 hours from 7:00 P.M., and you find the turkey must be placed in the oven no later than 1:00 P.M.

Example C   If your car gets 18.2 miles per gallon of gas, how many miles can you travel on 17.3 gallons of gas? Find your answer to the nearest tenth of a mile.

$$
\begin{array}{r}
18.2 \\
\times 17.3 \\
\hline
546 \\
12740 \\
18200 \\
\hline
314.86 \\
314.9
\end{array}
$$

18.2   Number of miles per gallon

$\times$ 17.3   Number of gallons

314.86   Mark off two decimal places, and round to nearest tenth

314.9   Number of miles per 17.3 gallons of gas to the nearest tenth

Example D   Gordon bought a fishnet for $11.79, and the sales tax is 4.5%. How much was the tax?

We need 4.5% of 11.79, and since 4.5% = 0.045, just compute 0.045 $\times$ 11.79.

0.045 $\times$ 11.79 = 0.53055

The tax will be $0.54. (When money is involved, the decimal is generally rounded up.)

Example E   Six children want to share $5 as equally as possible. How much should each receive?

Divide $5.00 by 6.

$$
\begin{array}{r}
\$0.83 \\
6\,\overline{)\$5.00} \\
-48 \\
\hline
20 \\
-18 \\
\hline
2
\end{array}
$$

2   Remainder is less than one-half of the divisor, quotient remains unchanged

Each child will receive $0.83 and there will be $0.02 left over.

A6.3 PROBLEMS

1. On a trip of 1076 miles, an automobile averaged 18.5 miles per gallon of gas. How much gasoline was used during this trip? Find the answer to the nearest tenth of a gallon. _____

2. If 3 pounds of potatoes cost $1.29, how much will 7 pounds of potatoes cost? $3.01 $1.29 ÷ 3 = .43

   $(.43)(7) = 3.01

3. Last summer, Lou earned $1250.85. She was paid $4.65 an hour. How many hours did she work? _____

4. A telephone company charges $1.95 for the first 3 minutes of usage and $0.21 for each additional minute. How much would a 15-minute phone call cost? $4.47

   $1.95 + 12(.21)$ $2.52 + 1.95 = 4.47

5. Pat needs 16.2 yards of material to make 9 blouses. To the nearest tenth of a yard, how much material is used to make one blouse? _____

6. If you need $18\frac{3}{4}$ pounds of a certain cut of meat to feed 15 people, how many pounds of the meat will you need to feed 10 people? _12½ lbs_

$$18\frac{3}{4} \div 15$$

$$5\ \frac{75}{4} \cdot \frac{1}{15} = \frac{5}{4}$$

$$\frac{5}{4} \cdot \frac{10}{1} = \frac{25}{2} = 12\frac{1}{2}\ lbs$$

7. Six water tanks of equal capacity hold 62,682 gallons of water. How many gallons will two tanks hold? _20,894_

$$\frac{62,682}{6} = 10,447\ gallons/tank$$

$$(10,447)(2) = 20,894$$

8. Using the same information as given in Problem 8, how many gallons of water will five tanks hold? _52,235_

$$62,682$$
$$10,447$$
$$52,235$$

9. A roast is to be cooked 18 minutes for each pound of weight. If a roast weighing 8 pounds is to be ready for dinner at 6:30 P.M., what time should it be put into the oven? _____

10. In 1987, the typewriter was 119 years old. In what year was the typewriter invented? __1868__

$$\begin{array}{r} 1987 \\ 119 \\ \hline 1868 \end{array}$$

11. A casserole was placed in the oven at 10:15 A.M. It must cook for 2 hours 30 minutes. When should it be removed from the oven? _____

12. The Smith family drove from El Paso, which is in the Mountain Time Zone, to Lubbock, which is in the Central Time Zone, in 8 hours 15 minutes. If they left El Paso at 7:30 A.M., what time was it in Lubbock when they arrived? __4:45pm__

$$\begin{array}{r} 7:30 \\ 8:15 \\ \hline 15:45 \end{array} = 3:45 + 1:00 = 4:45_p$$

13. For an engine repair job, a mechanic charges $42.50 plus $16.80 per hour. How much will the charge be for a job that takes six hours and 40 minutes? _____

14. A gift of jewelry cost Mr. Rios a total of $196.56, inclusive of a 5% sales tax. How much was the jewelry before the tax was added? _____

ELEMENTARY ALGEBRA SKILLS

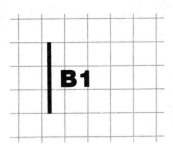

B1

OPERATIONS WITH POLYNOMIAL EXPRESSIONS

B1.1 SIMPLIFICATION OF A POLYNOMIAL BY GROUPING

In this topic, you will be working with expressions called polynomials. So we begin with telling what a polynomial is.

Arithmetic is concerned with numbers. In algebra, letters can represent numbers. Such letters are called **variables**.

Example A   Express the sum of the number 3 and the variable x.

It is written $3 + x$ or $x + 3$. Either way is all right.

Example B   Express the product of the number 3 and the variable x.

It is written $3x$, but not $x3$. The number comes first in such products.

Example C   Express the product of the number 3 and the variables s and t.

It is written $3st$. ($3ts$ is all right, but variables are generally written in alphabetical order.)

A **monomial** is a number, a variable, or a product of numbers and variables. A **polynomial** is a monomial or a sum of monomials. In such a sum, the monomials that make up the polynomial are called its **terms**. Polynomials are special cases of **algebraic expressions**.

Example D   Study and understand the following items.

1. $3 + x$ is a polynomial (but not a monomial). It has two terms, 3 and x.

2. $5 - 2r$ is a polynomial. Its terms are 5 and $-2r$.

3. -8 is a monomial and therefore a polynomial as well. All monomials are classified as polynomials.

4. $ab - 2a + 3b$ is a polynomial with the three terms ab, $-2a$, and $3b$.

5. $2mn - 4 + m - 6mn$ is a polynomial with the four terms $2mn$, -4, m, and $-6mn$.

6. In the monomial pq, the numerical multiplier (also called the numerical *coefficient*) is considered to be the number 1.

Grouping: In a polynomial, any terms that are exactly the same or that differ *only* in the numerical coefficient are called **like terms**. Such terms can be added by adding their numerical coefficients and leaving the variables the same. This is called **collecting like terms**, or **grouping**.

Example E   Write the polynomial $14a - 5a$ as a monomial.

Just as 14 days minus 5 days is 9 days, $14a - 5a = 9a$.

Example F   Collect like terms to simplify the polynomial $-4x + 9x - x$.

All three terms are like terms, so $-4x + 9x - x = 4x$.

Example G   Use grouping to simplify $3rs - 2m - rs + 8m$.

The terms $3rs$ and $-rs$ are like terms. They may be grouped to produce $2rs$. Also, $-2m$ and $8m$ are like terms. They may be grouped to produce $6m$. Thus, $3rs - 2m - rs + 8m = 2rs + 6m$. No further simplification is possible.

Example H   Simplify $4x - 2y + x - y$.

By grouping, the simplified form is $5x - 3y$.

Example I   Simplify $8ab - 2bc + ab - 4bc - 2ab$.

The simplified form, by grouping terms, is $7ab - 6bc$.

Example J   Simplify $3mn - 4m - 6mn + 5m - mn$.

The simplified form is $-4mn + m$. The custom is to write positive terms first where possible, so many people would write this as $m - 4mn$.

Example K $2mn - 4 + m - 6mn = m - 4mn - 4$. Do you recall this from Example D(5)?

B1.1 PROBLEMS

Write each as an algebraic expression.

1. The sum of the number 5 and the variable t._____

2. The sum of the number -8 and the variable n._____

3. The product of the number -4 and the variable x._____

4. The product of the number 9 and the variable k._____

5. The product of -2 and the variables r and s._____

6. The product of 5 and the variables p and q._____

7. The product of -7, $-x$, and y._____   9. The sum of $4u$ and $-5v$._____

8. The product of 6, m, and $-n$._____   10. The sum of $-8s$ and $-3t$._____

Write *yes* or *no* in each blank, as appropriate.

11. $8 - u$ 12. -7

   Monomial?_____   Monomial?_____

   Polynomial?_____   Polynomial?_____

13. 12   14. $1 + r$

   Monomial?_____   Monomial?_____

   Polynomial?_____   Polynomial?_____

15. $16r + 2 - 2$ 16. $-9pq$

   Monomial?_____   Monomial?_____

   Polynomial?_____   Polynomial?_____

17. $-14xy$

18. $5 - 2m + 3n$

Monomial?_____

Monomial?_____

Polynomial?_____

Polynomial?_____

Collect like terms to simplify each polynomial. Write each with the fewest number of terms.

19. $6x + 8x$_____

20. $9t - 2t =$_____

21. $3y - 5y =$_____

22. $5s + 11s =$_____

23. $7m - 5m + 2m =$_____

24. $8x + 6x - 9x =$_____

25. $-3s - 5s + s + 2s =$_____

26. $-4m + 6m - 2m - m =$_____

27. $6x - 2y - 2x =$_____

28. $5r + 3s - 3r =$_____

29. $-8m + 5 + 5m =$_____

30. $-6p - q + 2p =$_____

31. $5k - 4a - k =$_____

32. $-6n + 5b + n =$_____

33. $12s - 3t - 3s + 2t =$_____

34. $10x - 5y + 3x + 7y =$_____

35. $6a + 8b - 9a - 10b =$_____

36. $7m - 6b - 10m - 2n =$_____

37. $19rt + 2s + 3rt - s =$_____

38. $13k - 6ab - 9k + ab =$_____

39. $9a + 2b - a + b =$_____

40. $6x - y - x + 3y =$_____

41. $-14cd + 9c - cd - 3c =$_____

42. $-12p - 8pq + 9p + 9pq =$_____

43. $a - b + 2a - 2b - 3a =$_____

44. $-s + 4t + 2s - 5t + 3s =$_____

45. $\frac{2}{3}s + \frac{3}{4}t - s + \frac{1}{4}t =$_____

46. $\frac{5}{6}x - y - \frac{1}{2}x + \frac{1}{3}y =$_____

47. $\frac{a}{2} + b + \frac{a}{3} + \frac{b}{2} + \frac{a}{4} + \frac{b}{3} =$_____

48. $m - \frac{n}{2} + \frac{m}{2} - \frac{n}{3} + \frac{m}{3} - \frac{n}{4} =$_____

49. $3abc - 2ab + a - 2ab + 3abc =$_____

50. $x - 2xy + 3xyz - 2xy + x =$_____

B1.2 EVALUATION OF A POLYNOMIAL

Evaluating a polynomial is accomplished by substituting numbers for the variable or variables and then doing the arithmetic.

Example A Evaluate $x + 7$ when $x = 2$.

When $x = 2$, then $x + 7 = 2 + 7 = 9$. Thus, if $x = 2$, then $x + 7 = 9$.

Example B Evaluate $-2t + 3$ when $t = 4$.

If $t = 4$, then $-2t + 3 = -2(4) + 3 = -8 + 3 = -5$. Thus, if $t = 4$, then $-2t + 3 = -5$.

Example C Evaluate $4m - 2n$ when $m = 3$ and $n = 5$.

When $m = 3$ and $n = 5$, then $4m - 2n = 4(3) - 2(5) = 12 - 10 = 2$. So if $m = 3$ and $n = 5$, the value of $4m - 2n$ is 2.

Example D Find the value of the polynomial $5r + 3s - 17$ when $r = 2$ and $s = 2$.

Replacing r with 2 and replacing s with 2 gives

$$5r + 3s - 17 = 5(2) + 3(2) - 17$$
$$= 10 + 6 - 17$$
$$= 16 - 17 = -1$$

Example E Repeat Example D for the case that $r = -3$ and $s = 9$.

$$5r + 3s - 17 = 5(-3) + 3(9) - 17$$
$$= -15 + 27 - 17 = -5$$

Example F Repeat Example D for the case that $r = 4$ and $s = -1$.

$$5r + 3s - 17 = 5(4) + 3(-1) - 17$$
$$= 20 - 3 - 17 = 0$$

Example G Evaluate the polynomial $6mn + 5$ when $m = 2$ and $n = -2$.

When $m = 2$ and $n = -2$, then $6mn + 5 = 6(2)(-2) + 5 = -24 + 5 = -19$.

Example H Evaluate $5xy - 2y + 3x - 1$ when $x = -2$ and $y = -8$.

$$5xy - 2y + 3x - 1 = 5(-2)(-8) - 2(-8) + 3(-2) - 1$$
$$= 80 + 16 - 6 - 1 = 89$$

Example I Repeat Example H for $x = -8$ and $y = -2$.

$$5xy - 2y + 3x - 1 = 5(-8)(-2) - 2(-2) + 3(-8) - 1$$
$$= 80 + 4 - 24 - 1 = 59$$

Example J Evaluate $5xy - 2y + 3x - 3xy + x + 4y - 5$ when $x = 3$ and $y = -2$.

Before evaluating, you should collect like terms.

$$5xy - 2y + 3x - 3xy + x + 4y - 5 = 2xy + 4x + 2y - 5$$

Now there are only four terms, and

$$2xy + 4x + 2y - 5 = 2(3)(-2) + 4(3) + 2(-2) - 5$$
$$= -12 + 12 - 4 - 5 = -9$$

The next examples use exponents on variables. Just as $2^2 = 2 \cdot 2$ and $3^2 = 3 \cdot 3$, $x^2 = x \cdot x$ and $y^2 = y \cdot y$, and so on.

Example K Evaluate $x^2 - 3$ when $x = 4$.

$$x^2 - 3 = (4)^2 - 3$$
$$= 4 \cdot 4 - 3$$
$$= 16 - 3 = 13$$

Example L Evaluate $2a^2 - ab$ when $a = 3$ and $b = -2$.

$$2a^2 - ab = 2 \cdot 3^2 - 3(-2)$$
$$= 2 \cdot 9 - 3(-2)$$
$$= 18 + 6 = 24$$

Example M If $a = -4$ and $b = 2$, what is the value of $3a^2 - 4ab + 6b^2 + 30$?

$$3a^2 - 4ab + 6b^2 + 30 = 3(-4)^2 - 4(-4)(2) + 6 \cdot 2^2 + 30$$
$$= 3 \cdot 16 - 4(-4)(2) + 6 \cdot 4 + 30$$
$$= 48 + 32 + 24 + 30 = 134$$

Example N Evaluate $a^2b^2c - 2abc + 5a$ when $a = -3$, $b = 1$, and $c = -2$.

$$a^2b^2c - 2abc + 5a = (-3)^2(1)^2(-2) - 2(-3)(1)(-2) + 5(-3)$$
$$= 9(-2) - 2(-3)(-2) + 5(-3)$$
$$= -18 - 12 - 15 = -45$$

B1.2 PROBLEMS

Evaluate each polynomial for the indicated values of the variables.

1. If $x = 4$, then $x - 9 =$ _____

2. If $t = 3$, then $t + 7 =$ _____

3. If $s = 1$, then $5 - s =$ _____

4. If $r = 6$, then $9 - r =$ _____

5. If $m = 8$ and $n = -3$, then

$m + n =$ _____

6. If $d = 6$ and $e = 2$, then

$d - e =$ _____

7. If $p = 5$ and $q = 2$, then

$2p - q =$ _____

8. If $m = 2$ and $n = -2$, then

$3m + n =$ _____

9. If $c = 9$ and $d = 0$, then

$3c + 4d =$ _____

10. If $r = 6$ and $s = 3$, then

$5r - 3s =$ _____

11. If $a = 5$ and $b = 7$, then

$8a - 5b =$ _____

12. If $x = 0$ and $y = 8$, then

$5x + 2y =$ _____

13. If $s = 4$ and $t = 5$, then

$6s - 8t =$ _____

14. If $j = 5$ and $k = 9$, then

$9j - 5k =$ _____

15. If $y = 6$ and $k = 8$, then

$8y - 6k =$ _____

16. If $p = 5$ and $q = 6$, then

$7p - 9q =$ _____

17. If $m = -2$ and $n = -5$, then

$4m - 3n =$ _____

18. If $a = -4$ and $b = -5$, then

$-5a - 7b =$ _____

19. If $w = -7$ and $x = -6$, then

$-6w - 3x =$ _____

20. If $u = -3$ and $v = -4$, then

$3u - 4v =$ _____

21. If $r = -2$ and $s = 2$, then

$7r - 3s =$ _____

22. If $a = 4$ and $b = -2$, then

$8ab - a =$ _____

107

23. If $m = -2$ and $n = -3$, then

    $mn - 3n =$ _____

24. If $x = -3$ and $y = 1$, then

    $5x - 6y =$ _____

25. If $p = -4$ and $q = 2$, then

    $-5 - 3p + 2pq =$ _____

26. If $r = 3$ and $s = -2$, then

    $6r - 9 - 4rs =$ _____

27. If $a = 7$ and $b = -4$, then

    $3a - 2b - 5a + b =$ _____

28. If $s = -2$ and $t = 5$, then

    $5s + 3t - 8s - 7t =$ _____

29. If $x = -3$ and $y = 4$, then

    $8x - 9 + 2y - 3x =$ _____

30. If $m = 4$ and $n = -5$, then

    $7m - 3n - 8 - 5n =$ _____

31. If $p = 5$ and $q = -3$, then

    $-5p + 2 + p - 3q - 2p =$ _____

32. If $j = -5$ and $k = -4$, then

    $2jk + 2k - 5jk - 6k =$ _____

33. If $f = -6$ and $g = -3$, then

    $6f - 4fg - 5 + 2f - 5fg - 3g =$ _____

34. If $c = -4$ and $d = 3$, then

    $-4c - 2d - c + 5 + 3c =$ _____

35. If $t = 3$, then $t^2 - 6 =$ _____

36. If $k = 3$, then $2k^2 - k - 1 =$ _____

37. If $n = -2$, then $5n^2 + 6n - 3 =$ _____

38. If $x = -4$, then $x^2 + 9 =$ _____

39. If $w = 3$, $x = -5$, then $6w^2 + wx - 1 =$ _____

40. If $s = 2$, $t = -3$, then $-5s + 4t^2 - st =$ _____

41. If $x = 4$, $y = -4$, then $-5x + x^2y - 7 + 3x =$ _____

42. If $a = -3$, $b = -3$, $c = 2$, then $7abc - 2a^2b + 4 =$ _____

43. If $p = -2$, $q = -2$, $r = 3$, then $9 - 2pqr - 3pq^2 =$ _____

44. If $a = -5$, $b = -1$, then $-6a^2b + 2a + 3 - 5a =$ _____

45. If $a = -2$, $b = -2$, then $3a^2 - 2b^2 + 4ab + 1 =$ _____

46. If $x = -3$, $y = -3$, then $2x^2 - 3xy + 5y^2 - 7 =$ _____

47. If $a = 2$, $b = 2$, then $3a^2 - 2b^2 + 4ab + 1 =$ _____

48. If $x = 3$, $y = 3$, then $2x^2 - 3xy + 5y^2 - 7 =$ _____

B1.3 ADDITION AND SUBTRACTION OF POLYNOMIALS

In this unit, you will add and subtract polynomials. You have already done some of this in B1.1. The idea is simply to collect like terms.

Example A   Find the sum of $2x^2 + 3x + 7$ and $4x^2 + 2x + 5$.

We can restate the problem as: Find $(2x^2 + 3x + 7) + (4x^2 + 2x + 5)$. Now $2x^2$ and $4x^2$ are like terms. And just as 2 days plus 4 days is 6 days, so $2x^2 + 4x^2 = 6x^2$. Likewise, $3x$ and $2x$ are like terms. Their sum is $5x$. Finally, $7 + 5 = 12$. Thus,

$$(2x^2 + 3x + 7) + (4x^2 + 2x + 5) = 6x^2 + 5x + 12$$

Another way to do this is through a "vertical" method of writing the polynomials to be added. You line up the like terms. It goes like this:

$$
\begin{array}{r}
2x^2 + 3x + 7 \\
4x^2 + 2x + 5 \\
\hline
6x^2 + 5x + 12
\end{array}
$$

Example B   Find $(5x^2 + 7x + 8) + (3x^2 - 2x - 1)$.

Add like terms. The sum is $8x^2 + 5x + 7$. Using the vertical method,

$$
\begin{array}{r}
5x^2 + 7x + 8 \\
3x^2 - 2x - 1 \\
\hline
8x^2 + 5x + 7
\end{array}
$$

Example C   Find $(8x^2 + 2x + 3) + (x^2 + 2x - 6)$.

Add like terms and remember that x^2 means $1x^2$. The sum is $9x^2 + 4x - 3$. Check this with the vertical method.

$$
\begin{array}{r}
8x^2 + 2x + 3 \\
x^2 + 2x - 6 \\
\hline
9x^2 + 4x - 3
\end{array}
$$

Example D   Find $(3y^2 + 2y - 5) + (y^2 - 5y + 1)$.

Adding like terms gives $4y^2 - 3y - 4$. Again, check by the vertical method.

$$
\begin{array}{r}
3y^2 + 2y - 5 \\
y^2 - 5y + 1 \\
\hline
4y^2 - 3y - 4
\end{array}
$$

Example E   Find $(-2t^2 + 3t - 4) + (t^2 - 5t + 4)$.

Combine like terms or add vertically. You get $-t^2 - 2t$.

109

Example F   Find $(3t^2 + 4t - 9) + (-3t^2 - t + 2)$.

This time, the t^2 term vanishes and the sum is $3t - 7$.

Example G   Subtract $5u^2 - 2u + 3$ from $3u^2 + 2u - 1$.

This time we seek $(3u^2 + 2u - 1) - (5u^2 - 2u + 3)$. It looks different because of the subtraction involved. But subtracting $5u^2 - 2u + 3$ is just the same as adding $-5u^2 + 2u - 3$. Thus,

$$(3u^2 + 2u - 1) - (5u^2 - 2u + 3) = (3u^2 + 2u - 1) + (-5u^2 + 2u - 3)$$

$$= -2u^2 + 4u - 4$$

Rule: To subtract one polynomial from another, change the sign of each term being subtracted, then add.

Example H   Find $(8k^2 - 2k + 3) - (2k^2 - 5k - 6)$.

$$(8k^2 - 2k + 3) - (2k^2 - 5k - 6) = (8k^2 - 2k + 3) + (-2k^2 + 5k + 6)$$

$$= 6k^2 + 3k + 9$$

Example I   Find $(-3s^2 + 4s - 5) - (5s^2 - 2s + 7)$.

$$(-3s^2 + 4s - 5) - (5s^2 - 2s + 7) = (-3s^2 + 4s - 5) + (-5s^2 + 2s - 7)$$

$$= -8s^2 + 6s - 12$$

The goal here is to add or subtract polynomials without taking too many intermediate steps. You may want to rearrange terms and to add or subtract like terms in your head.

Example J   Find $(4x^2 - 3 + 5x) - (-x - x^2 + 6)$.

Taking intermediate steps, we proceed as follows:

$$(4x^2 - 3 + 5x) - (-x - x^2 + 6) = (4x^2 + 5x - 3) - (-x^2 - x + 6)$$

$$= (4x^2 + 5x - 3) + (x^2 + x - 6)$$

$$= 5x^2 + 6x - 9$$

See if you could have done it without intermediate steps.

Example K   Find $(7 - 2x^2 + x) - (5x - 3 + 4x^2)$, without taking intermediate steps.

The answer is $-6x^2 - 4x + 10$. Were you able to do it? If you need, check the answer by use of the intermediate steps.

B1.3 PROBLEMS

Compute each sum or difference, as indicated.

1. $(3x^2 + 2x + 8) + (5x^2 - 3x + 1) =$ _$8x^2 - x + 9$_

2. $(4x^2 + 4x + 5) + (2x^2 + 2x + 4) =$ _$6x^2 + 6x + 9$_

3. $(7x^2 + 5x + 3) + (2x^2 - 2x - 1) =$ _$9x^2 + 3x + 2$_

4. $(6x^2 + 6x + 7) + (3x^2 - 4x - 3) =$ _$9x^2 + 2x + 4$_

5. $(4y^2 + y + 3) + (2y^2 + 2y - 1) =$ _____

6. $(y^2 + 3y + 2) + (3y^2 - 2y + 2) =$ _____

7. $(u^2 + 3u - 7) + (2u^2 - 6u + 3) =$ _____

8. $(3u^2 - 5u + 4) + (u^2 + 3u - 6) =$ _____

9. $(2v^2 - 3v + 2) + (v^2 + 3v - 3) =$ _____

10. $(v^2 + 5v - 3) + (3v^2 - 4v + 3) =$ _____

11. $(5w^2 - 2w + 4) + (2w^2 + w - 4) =$ _____

12. $(4w^2 + 5w - 3) + (3w^2 - 5w + 5) =$ _____

13. $(6t^2 - 4t + 5) + (-6t^2 + 4t - 3) =$ _____

14. $(-t^2 + 3t - 4) + (t^2 - 2t + 4) =$ _____

15. $(m - 5 + 2m^2) + (5 - 2m^2 - m) =$ _____

16. $(3 - 3m^2 - 4m) + (3m^2 - 3 + 4m) =$ _____

17. $(6x^2 - 5x + 4) - (3x^2 + 2x - 1) =$ _____

18. $(4x^2 + 2x - 5) - (2x^2 - 2x + 3) =$ _____

19. $(9x^2 - 7x - 7) - (6x^2 - x + 2) =$ _____

20. $(7x^2 - 3x + 4) - (3x^2 - 5x - 4) =$ _____

111

21. $(t^2 - 4t + 6) - (3t^2 + 5t - 5) =$ _____

22. $(2t^2 - 5t + 2) - (t^2 - 2t + 3) =$ _____

23. $(r^2 + r - 5) - (-r^2 - r - 5) =$ _____

24. $(r^2 - r + 6) - (-r^2 + r + 6) =$ _____

25. $(2a^2 + 5 - a) - (-a^2 - 2a + 3) =$ _____

26. $(a^2 - 6 + 2a) - (-3a^2 + 4a - 4) =$ _____

27. $(4m - 2m^2 - 3) - (6m - 5 + 3m^2) =$ _____

28. $(8 - 5m^2 + 6m) - (-4m - 7 + 2m^2) =$ _____

29. $(9 - 4x + 5x^2) - (-10 - 2x^2 + 3x) =$ _____

30. $(5x - 3x^2 - 7) - (5 - 3x + 5x^2) =$ _____

31. $(3 + 4t - 5t^2 - 5 - t) + (2t - 7 - 4t^2 + t) =$ _____

32. $(6t - 8 + 4t^2 - 3) + (9 - 4t^2 + 2t - t^2 + 5) =$ _____

33. $(5 - k^3 + 2k - 4k^2) - (-3k + 2k^2 - 7 - 3k^3) =$ _____

34. $(6k^2 - 4 + 3k^3 - 8k) - (-7 - 2k^3 + k^2 - 5k) =$ _____

35. $(8u - 3 + 6u^2 + 8 - 4u^3) - (2u^2 - 7u + 9 - 2u^3) =$ _____

36. $(4u^2 - 6u^3 - 7u + 9) - (12u - 3u^3 + 4 - 5u - 3u^2) =$ _____

37. $(6 - 5v - 2v^3 + 8 - 6v) - (4v^2 - 9 + 5v^3 + 3 + 5v^2 - v^3) =$ _____

38. $(4v^2 - 3v^3 + 8 - 9v^2 - 11) - (v^3 - 2v - 6 + 3v^3 + 8 + 5v) =$ _____

39. $(12x^3 - 10 + 5x^2 - 3x - 5 - 8x) - (2x^2 + 6x^3 - 4x + 3x^2 + 6x^3 - 3) =$ _____

40. $(13 - 7x^2 + 3x^3 - 5x - 20 - 8x^3) - (9x - 2x^3 - 13 - 7x^2 - 3x^3 - 12) =$ _____

B1.4 MULTIPLICATION OF A MONOMIAL WITH A POLYNOMIAL

You have already learned about numbers with positive integer exponents. Variables with positive integer exponents work the same way; after all, variables are just letters that represent numbers. Thus,

x^1 means x
x^2 means $x \cdot x$
x^3 means $x \cdot x \cdot x$
x^4 means $x \cdot x \cdot x \cdot x$

and so on.

Example A Write the product $x^3 \cdot x^2$ in a simpler way.

Writing $x^3 = x \cdot x \cdot x$ and $x^2 = x \cdot x$, then

$$x^3 \cdot x^2 = (x \cdot x \cdot x)(x \cdot x)$$
$$= x \cdot x \cdot x \cdot x \cdot x$$
$$= x^5$$

Example B Find the product $(2x^2y)(3x^3y^2)$.

Taking things one at a time. First, $2 \cdot 3 = 6$. Next, $x^2 \cdot x^3 = x^5$, as you saw in Example A. Finally, $y \cdot y^2 = y \cdot y \cdot y = y^3$. Thus, the product is $6x^5y^3$.

Example C Find the product $(a^2b)(4ab)$.

Remember that a^2b means $1 \cdot a^2b$. In order, $1 \cdot 4 = 4$; $a^2 \cdot a = a \cdot a \cdot a = a^3$; and $b \cdot b = b^2$. Therefore, $(a^2b)(4ab) = 4a^3b^2$.

Example D Verify the result in Example C for the case $a = 2$ and $b = 3$.

If $a = 2$ and $b = 3$, then $a^2b = 2^2 \cdot 3 = 4 \cdot 3 = 12$. Also, $4ab = 4 \cdot 2 \cdot 3 = 24$. Thus,

$$(a^2b)(4ab) = 12 \cdot 24$$
$$= 288$$

Now we evaluate $4a^3b^2$ for the case $a = 2$ and $b = 3$.

$$4a^3b^2 = 4 \cdot 2^3 \cdot 3^2$$
$$= 4 \cdot 8 \cdot 9$$
$$= 32 \cdot 9$$
$$= 288$$

113

We've been multiplying monomials by monomials. We now multiply monomials by polynomials, using the distributive property: $a(b + c) = ab + ac$.

Example E   Write $x^2(2x + 3)$ as a polynomial.

$$x^2(2x + 3) = x^2(2x) + x^2(3)$$

$$= 2x^3 + 3x^2$$

Example F   Find the product $3a^2(2a^2 - 3a)$.

Again, multiply each term of the second factor by $3a^2$, as follows:

$$3a^2(2a^2 - 3a) = 3a^2 \cdot 2a^2 - 3a^2 \cdot 3a$$

$$= 6a^4 - 9a^3$$

Example G   Find $2a^3(4a^2 + 6a - 3)$.

$$2a^3(4a^2 + 6a - 3) = 2a^3 \cdot 4a^2 + 2a^3 \cdot 6a + 2a^3 \cdot (-3)$$

$$= 8a^5 + 12a^4 - 6a^3$$

Example H   Find $2x^2y(3x^3y^2 - 4xy)$.

Notice that the first part of this is just like Example B.

$$2x^2y(3x^3y^2 - 4xy) = (2x^2y)(3x^3y^2) - (2x^2y)(4xy)$$

$$= 6x^5y^3 - 8x^3y^2$$

Example I   Find $-5ab^2(2a^2 - 3ab - 2b)$.

Don't let the minus signs give you trouble on this one.

$$-5ab^2(2a^2 - 3ab - 2b)$$

$$= (-5ab^2)(2a^2) - (-5ab^2)(3ab) - (-5ab^2)(2b)$$

$$= -10a^3b^2 + 15a^2b^3 + 10ab^3$$

Example J   Find $7u^3v^2w(-8uv + 5uw - 3vw)$ without intermediate steps and see if you can do it in your head.

The result is

$$-56u^4v^3w + 35u^4v^2w^2 - 21u^3v^3w^2$$

You save a lot of time if you develop your skills to the point of doing things in your head and getting the right answers.

B1.4 PROBLEMS

Perform the indicated multiplications and verifications.

1. $x^3 \cdot x =$ _____ x^4

2. $a^4 \cdot a^2 =$ _____ a^6

3. $b^2 \cdot b^2 =$ _____ b^4

4. $c^3 \cdot c^3 =$ _____ c^6

5. $(3u^2)(4uv) =$ _____ $12u^3v$

6. $(3mn)(-3m^3) =$ _____ $-9m^4n$

7. $(-2ab)(6a^2) =$ _____ $-12a^3b$

8. $(5s^3)(3st) =$ _____ $15s^4t$

9. $(-7c^2d^2)(-4c^3d) =$ _____ $28c^5d^3$

10. $(-8p^2q^3)(3pq^2) =$ _____ $-24p^3q^5$

11. Verify your answer to Problem 5 for the case $u = 2$, $v = 3$.

12. Verify your answer to Problem 6 for the case $m = 2$, $n = 3$.

13. Verify your answer to Problem 7 for the case $a = 3$, $b = 2$.

14. Verify your answer to Problem 8 for the case $s = 3$, $t = 2$.

15. Verify the result in Example E for the case $x = 3$.

16. Verify the result in Example F for the case $a = 4$.

17. $s^3(3s + 2) =$ _____ $3s^4 + 2s^3$

18. $x^4(2x + 5) =$ _____ $2x^5 + 5x^4$

19. $3n(n^2 - 2) =$ _____ $3n^3 - 6n$

20. $2r(3r - 4) =$ _____ $6r^2 - 8r$

21. $5s^2(2s - 3) =$ _____ $10s^3 - 15s^2$

22. $4k^2(4k - 3) =$ _____ $16k^3 - 12k^2$

23. $6m^2(3m^2 - 2m) =$ _____ $18m^4 - 12m^3$

24. $7bc(b^2 + 2c^2) =$ $7b^3c + 14bc^3$

25. $2c^3(xy + 4y) =$ $2c^3xy + 8c^3y$

26. $5t^3(3t - 4t^2) =$ $15t^4 - 20t^5$

27. $9a^4(3a^2 - 2a + 4) =$ $27a^6 - 18a^5 + 36a^4$

28. $-6b^2c(-3bc - 2c^2 + 4) =$ $9b^3c^2 + 12b^2c^3 - 24b^2c$

29. $-7pq^2(2p^2q - 4p - 2) =$ $-14p^3q^3 + 28p^2q^2 + 14pq^2$

30. $8t^5(4t^3 - 3t - 5) =$ $32t^8 - 24t^6 - 40t^5$

31. $5g^3h^2(-4g^2h - 2g + h) =$ $-20g^5h^3 - 10g^4h^2 + 5g^3h^3$

32. $9x^2y^4(2x^3 - y - 6xy^2) =$ _____

33. $-8a^3b^2c(-6abc^2 - 5b^3 - 4c^2) =$ _____

34. $-7pq^2r^4(-5p^2q - 7pq^3r - 4r) =$ _____

35. $6u^4v^3(-5u^3v + 7uv^2 - 9w^2) =$ _____

36. $5m^5n^2(-7m^2n^3 - 3m^4 + 5p^3) =$ _____

37. $-8x^4y^5z^2(-8x^5y^4z^3 - 9x + 7y^2 - 3x^5) =$ _____

38. $-4a^6b^4c^3(9a^3b^4c^5 - 7a^2 - 8b^3 - c^6) =$ _____

39. $3u^7v^8(12u^5v^3w^2 - 15u^7w^3 - 18v^8 + 17w^{10}) =$ _____

40. $4g^6h^9(-13g^8h^5k^3 - 19h^4k^2 + 16g^9k^7 - 18k^{12}) =$ _____

B1.5 MULTIPLICATION OF TWO BINOMIALS

You have learned that a polynomial is a sum of monomials. If a polynomial has only two terms, it is called a **binomial**. If a polynomial has only three terms, it is called a **trinomial**. We do not use special names for polynomials with more than three terms.

Multiplying a polynomial by a monomial does not increase the number of terms. Multiplying two binomials, however, can result in a polynomial with as many as four terms or as few as two terms.

Example A   Find the product of $x + 3$ and $2x + 1$ by using the distributive property.

Write this as $(x + 3)(2x + 1)$. Multiply each term of $2x + 1$ by $x + 3$:

$$(x + 3)(2x + 1) = (x + 3)(2x) + (x + 3)(1)$$

In this way, the binomial $x + 3$ now gets multiplied by a monomial. Let's continue.

$$(x + 3)(2x + 1) = (x + 3)(2x) + (x + 3)(1)$$
$$= 2x(x + 3) + (x + 3)$$
$$= 2x^2 + 6x + x + 3$$
$$= 2x^2 + 7x + 3$$

Notice that the product at first contained four terms. But two of them, the $6x$ and the x, are like terms that can be combined. The result is a trinomial.

Example B   Find $(x + 3)(2x + 1)$ in another way.

This time, we multiply each term of $x + 3$ by $2x + 1$:

$$(x + 3)(2x + 1) = x(2x + 1) + 3(2x + 1)$$
$$= 2x^2 + x + 6x + 3$$
$$= 2x^2 + 7x + 3$$

Example C   Verify the results of Examples A and B for the case $x = 5$.

When $x = 5$, then   Also, when $x = 5$, then

$(x + 3)(2x + 1) = (5 + 3)(10 + 1)$ $2x^2 + 7x + 3 = 2 \cdot 5^2 + 7 \cdot 5 + 3$

$\qquad\qquad\qquad = 8 \cdot 11$ $\qquad\qquad\qquad = 2 \cdot 25 + 35 + 3$

$\qquad\qquad\qquad = 88$ $\qquad\qquad\qquad = 50 + 38$

$\qquad\qquad\qquad\qquad\qquad\qquad = 88$

117

Example D   Find $(2x + 3)(3x + 2)$.

Here's one way to do it:

$$(2x + 3)(3x + 2) = 2x(3x + 2) + 3(3x + 2)$$

$$= 6x^2 + 4x + 9x + 6$$

$$= 6x^2 + 13x + 6$$

There is a general method for multiplying two binomials. To develop the method, let $A + B$ represent any binomial and let $C + D$ represent any binomial. Now study this next example.

Example E   Find $(A + B)(C + D)$.

We proceed as before with the distributive property.

$$(A + B)(C + D) = A(C + D) + B(C + D)$$

$$= AC + AD + BC + BD$$

Notice the result.

$$(A + B)(C + D) = AC \qquad + AD \qquad + BC \qquad + BD$$

	↑	↑	↑	↑
	Product of the First terms	Product of the Outer terms	Product of the Inner terms	Product of the Last terms

So you only need to remember First, Outer, Inner, and Last. In shorthand, remember FOIL. This is called the FOIL method for multiplying two binomials.

Example F   Use the FOIL method to compute the product $(a + x)(y + 2)$.

$$(a + x)(y + 2) = ay \quad + 2a \quad + xy \quad + 2x$$

	↑	↑	↑	↑
	First	Outer	Inner	Last

Example G   Use FOIL to find $(2r - 5)(x + 4)$.

$$(2r - 5)(x + 4) = 2rx \quad + 8r \quad - 5x \quad - 20$$

	↑	↑	↑	↑
	F	O	I	L

Example H   Find $(2x + 3)(2x - 3)$.

$$(2x + 3)(2x - 3) = 4x^2 - 6x + 6x - 9$$
$$= 4x^2 - 9$$

Example I   Find $(3a + 4b)(3a - 4b)$.

$$(3a + 4b)(3a - 4b) = 9a^2 - 12ab + 12ab - 16b^2$$
$$= 9a^2 - 16b^2$$

Example J   Find $(3x^2 - 2)(2x - 1)$.

Use FOIL again.

$$(3x^2 - 2)(2x - 1) = 6x^3 - 3x^2 - 4x + 2$$

Example K   Find $(2x - 3y)(2x + y)$.

$$(2x - 3y)(2x + y) = 4x^2 + 2xy - 6xy - 3y^2$$
$$= 4x^2 - 4xy - 3y^2$$

Example L   Find $(2x - 3y)(2x + 3y)$.

$$(2x - 3y)(2x + 3y) = 4x^2 + 6xy - 6xy - 9y^2$$
$$= 4x^2 - 9y^2$$

Example M   Find $(r + 4s)(3r - 4s)$.

$$(r + 4s)(3r - 4s) = 3r^2 - 4rs + 12rs - 16s^2$$
$$= 3r^2 + 8rs - 16s^2$$

Example N   Find $(1 - 2k)(2k - 1)$.

$$(1 - 2k)(2k - 1) = 2k - 1 - 4k^2 + 2k$$
$$= -4k^2 + 4k - 1$$

Example O   Find $(u^2 - 2v)(u + v)$.

$$(u^2 - 2v)(u + v) = u^3 + u^2v - 2uv - 2v^2$$

Example P   Find $(5a^2 - b^2)(a + b)$.

$$(5a^2 - b^2)(a + b) = 5a^3 + 5a^2b - ab^2 - b^3$$

Example Q   Find $(4x^2y + 1)(2xy^2 - 1)$.

$$(4x^2y + 1)(2xy^2 - 1) = 8x^3y^3 - 4x^2y + 2xy^2 - 1$$

Example R   Find $(6xy - z)(5xz^2 + y^2)$.

$$(6xy - z)(5xz^2 + y^2) = 30x^2yz^2 + 6xy^3 - 5xz^3 - y^2z$$

B1.5 PROBLEMS

Multiply these binomials. Collect like terms whenever possible to simplify your answers.

1. $(x + 4)(x + 3) =$ $x^2 + 7x + 12$

2. $(y + 2)(y + 5) =$ $y^2 + 7y + 10$

3. $(2y + 7)(y + 6) =$ $2y^3 + 13y + 13$

4. $(x + 3)(3x + 4) =$ $3x^2 + 13x + 12$

5. $(2a + b)(a - 3b) =$ $2a^2 - 5ab - 3b^2$

6. $(m - 4n)(3m + 2n) =$ $3m^2 - 10mn - 8n^2$

7. $(2p - 3q)(3p - 2q) =$ $6p^2 - 13pq + 6q^2$

8. $(3x - 5y)(5x - 3y) =$ $15x^2 - 34xy + 5y^2$

9. $(4b - c)(4b + c) =$ _____

10. $(3j + 2k)(3j - 2k) =$ _____

11. $(5x + 4y)(3x + 5y) =$ _____

12. $(7a + 3b)(5a + 2b) =$ _____

13. $(2s - 3t)(3t - 2s) =$ _____

14. $(4u - 5v)(5v - 4u) =$ _____

15. $(2ab + c)(3ab - 2c) =$ _____

16. $(3r - 4st)(2r + st) =$ _____

17. $(4 + 3n)(5 - 2n) =$ _____

18. $(6 - 5c)(7 + 3c) =$ _____

19. $(2v + w)(2v - w) =$ _____

20. $(5s - t)(5s + t) =$ $25s^2 - t^2$

121

21. $(7x - 2y)(7x + 2y) = $ _$49x^2 - 4y^2$_

22. $(4a + 5b)(4a - 5b) = $ _$16a^2 - 25b^2$_

23. $(5m - 6n)(6n + 5m) = $ _$5m^2 - 36n^2$_

24. $(3q + 7p)(7p - 3q) = $ _$-9q^2 + 42pq + 49p^2$_

25. $(2a + 3a)(4b + b) = $ _____

26. $(x + 3x)(5y + 2y) = $ _____

27. $(a^2 - b)(b + a^2) = $ _____

28. $(m^2 + n^3)(n^3 - m^2) = $ _____

29. $(2x + y)(3x + z) = $ _____

30. $(p + 3q)(r + 5q) = $ _____

31. $(2r^2 - 3s^2)(r + s) = $ _____

32. $(3a + 2b)(a^2 - b^2) = $ _____

33. $(7uv - w)(3w + uv) = $ _____

34. $(5j + 3jk)(4jk - j) = $ _____

35. $(3r^2s - 2t^2)(3r^2s + 2t^2) = $ _____

36. $(5m^3 + 2uv^2)(5m^3 - 2uv^2) = $ _____

37. $(8wx^2 + 9w^2x)(3w^3 - 2x^3) = $ _____

38. $(2p^2 + 5q^2)(7pq^3 - 8p^3q) = $ _____

39. $(5a^4bc + 3ab^3)(6ac^2 - 5b^2c) = $ _____

40. $(8x^2y^3 - 9xyz^2)(3x^3y + 2y^2z) = $ _____

B1.6 SQUARING A BINOMIAL

In this unit, you will learn to multiply a binomial by itself (in other words, to find the square of a given binomial). The FOIL method will lead to a general rule for squaring binomials.

Example A Use the FOIL method to find $(x + 7)^2$.

$(x + 7)^2$ means $(x + 7)(x + 7)$. Now by the FOIL method,

$$(x + 7)(x + 7) = x^2 + 7x + 7x + 7^2$$
$$\underset{F \quad\quad O \quad\quad I \quad\quad L}{\uparrow \quad\quad \uparrow \quad\quad \uparrow \quad\quad \uparrow}$$
$$= x^2 + 14x + 49$$

Example B Use FOIL to find $(n + 9)^2$.

$$(n + 9)^2 = (n + 9)(n + 9)$$
$$= n^2 + 9n + 9n + 81$$
$$= n^2 + 18n + 81$$

Example C Find $(r - 5)^2$.

By FOIL again,

$$(r - 5)^2 = (r - 5)(r - 5)$$
$$= r^2 - 5r - 5r + 25$$
$$= r^2 - 10r + 25$$

There is a general rule which you may have already noticed from the three examples presented so far. To this end, let $A + B$ represent a general binomial.

Example D Find $(A + B)^2$.

Use FOIL in the usual way.

$$(A + B)^2 = (A + B)(A + B)$$
$$= A^2 + AB + BA + B^2$$
$$= A^2 + 2AB + B^2$$

> RULE FOR SQUARING A BINOMIAL
> The square of a binomial is the square of the first term, plus twice the product of the two terms, plus the square of the second term.

See Examples A, B, and C for verification of this rule.

123

Example E   Use the rule to find $(2r + 3s)^2$.

$$(2r + 3s)^2 = (2r)^2 \quad + 2(2r)(3s) \quad + (3s)^2$$

$$\qquad\qquad \uparrow \qquad\qquad \uparrow \qquad\qquad \uparrow$$

Square	Twice the	Square
of the	product of	of the
first term	the two terms	second term

$$= 4r^2 + 12rs + 9s^2$$

Example F   Use the rule to find $(5m - 3n)^2$.

$$(5m - 3n)^2 = (5m)^2 + 2(5m)(-3n) + (-3n)^2$$

$$= 25m^2 - 30mn + 9n^2$$

Example G   Use the rule to find $(3ab - 2a)^2$.

$$(3ab - 2a)^2 = (3ab)^2 + 2(3ab)(-2a) + (-2a)^2 = 9a^2b^2 - 12a^2b + 4a^2$$

Example H   Find $(3x + 1)^2$.

$$(3x + 1)^2 = (3x)^2 + 2(3x)(1) + 1^2 = 9x^2 + 6x + 1$$

Example I   Find $(a^2 - b^2)^2$.

$$(a^2 - b^2) = a^4 - 2a^2b^2 + b^4$$

Example J   Find $(5a^3b - 2ab^2)^2$.

$$(5a^3b - 2ab^2)^2 = 25a^6b^2 - 20a^4b^3 + 4a^2b^4$$

Example K   Find $\left(\dfrac{a}{2} + b\right)^2$.

$$\left(\frac{a}{2} + b\right)^2 = \left(\frac{a}{2}\right)^2 + 2\left(\frac{a}{2}\right)(b) + b^2 = \frac{a^2}{4} + ab + b^2$$

Example L   Find $\left(n + \dfrac{1}{2}\right)^2$.

$$\left(n + \frac{1}{2}\right)^2 = n^2 + n + \frac{1}{4}$$

This last example can be used to perform a numerical mental trick: squaring a mixed number whose fractional part is $\dfrac{1}{2}$. Look at this:

$$\left(2\frac{1}{2}\right)^2 = \left(2 + \frac{1}{2}\right)^2 = 4 + 2 + \frac{1}{4} = 6\frac{1}{4}$$

Example M $\left(5\dfrac{1}{2}\right)^2 = 30\dfrac{1}{4}.$ **Example N** $\left(7\dfrac{1}{2}\right)^2 = 56\dfrac{1}{4}.$

B1.6 PROBLEMS

Find each square.

1. $(r + 2)^2 =$ _____ $r^2 + 4r + 4$ _____

2. $(s - 3)^2 =$ _____ $s^2 + 6s + 9$ _____

3. $(t - 4)^2 =$ _____ $t^2 + 8t + 16$ _____

4. $(u + 5)^2 =$ _____ $u^2 + 10u + 25$ _____

5. $(3 - s)^2 =$ _____

6. $(4 - t)^2 =$ _____

7. $(x + y)^2 =$ _____

8. $(u - v)^2 =$ _____

9. $(2a - 5b)^2 =$ _____

10. $(4u - 3v)^2 =$ _____

11. $(5 - 2rs)^2 =$ _____

12. $(7pq - 3)^2 =$ _____

13. $(8a^2 + b)^2 =$ _____

14. $(7m + n^2)^2 =$ _____

15. $(2s + 3t)^2 =$ _____

16. $(3r + 2s)^2 =$ _____

17. $(4pq + 3r)^2 =$ _____

18. $(5a + 6bc)^2 =$ _____

19. $\left(\dfrac{1}{3} + 3x\right)^2 =$ _____

20. $\left(4t + \dfrac{1}{4}\right)^2 =$ _____

125

21. $\left(\dfrac{1}{2} - 2r\right)^2 =$ _____

22. $\left(3s - \dfrac{1}{2}\right)^2 =$ _____

23. $\left(a + \dfrac{1}{a}\right)^2 =$ _____

24. $\left(\dfrac{1}{x} - x\right)^2 =$ _____

25. $(4m^2n - 3m)^2 =$ _____

26. $(2ab^3 - 5a^2)^2 =$ _____

27. $(6m^3 - m^2n)^2 =$ _____

28. $(pq^3 - 5p^2)^2 =$ _____

29. $(2u^2v^3 + 5v^2w)^2 =$ _____

30. $(4s^3t + 3s^2t^4)^2 =$ _____

31. $(7ab^2c^3 - 8a^3b^2c)^2 =$ _____

32. $(9u^3v^2w - 5uv^2w^3)^2 =$ _____

33. $\left(5j + \dfrac{3kv}{4}\right)^2 =$ _____

34. $\left(\dfrac{2rs}{3} + 4t\right)^2 =$ _____

35. $\left(3\dfrac{1}{2}\right)^2 = \underline{3^2 + 2 + 1/2^2 = 9 + 2 + \frac{1}{4} = 11\frac{1}{4}}$

36. $\left(6\dfrac{1}{2}\right)^2 = \underline{6^2 + 2 + \frac{1}{2}^2 = 36 + 2 + \frac{1}{4} = 38\frac{1}{4}}$

37. $(9.5)^2 = \underline{83\frac{1}{4}}$

38. $(8.5)^2 = \underline{66\frac{1}{4}}$

Write each as simply as you can.

39. $(x + y)^2 + (x - y)^2 =$ _____

40. $(x + y)^2 - (x - y)^2 =$ _____

B1.7 DIVISION OF POLYNOMIALS WITH A MONOMIAL DIVISOR

You have been multiplying polynomials. We now approach the subject of dividing polynomials. In this unit, we restrict ourselves to the case of a monomial divisor. The method we will use resembles long division for numbers.

Example A Find $\dfrac{9x + 6}{3}$.

Set this up as $3\overline{)9x + 6}$. Then write $3x$ above the $9x$ since $3 \cdot 3x = 9x$.

$$
\begin{array}{r}
3x \\
3\overline{)9x + 6}
\end{array}
$$

As with numbers, multiply and subtract.

$$
\begin{array}{r}
3x \\
3\overline{)9x + 6} \\
(-)\underline{9x}
\end{array}
$$

Bring down the 6 and divide 6 by 3, putting the 2 above the 6.

$$
\begin{array}{r}
3x + 2 \\
3\overline{)9x + 6} \\
\underline{9x} \\
6 \\
(-)\underline{6}
\end{array}
$$

Again, multiply and subtract to complete the process and to check your division. The final form looks like this:

$$
\begin{array}{r}
3x + 2 \\
3\overline{)9x + 6} \\
\underline{9x} \\
6 \\
\underline{6} \\
0
\end{array}
$$

As with numbers, you can check that this division is correct by multiplying: $3(3x + 2) = 9x + 6$. Thus $\dfrac{9x + 9}{3} = 3x + 2$.

Example B Find $\dfrac{18y - 72}{6}$.

$$
\begin{array}{r}
3y - 12 \\
6\overline{)18y - 72} \\
\underline{18y} \\
-72 \\
\underline{-72} \\
0
\end{array}
$$

Again, we check that the result is correct by multiplying: $6(3y - 12) = 18y - 72$. Thus, $\dfrac{18y - 72}{6} = 3y - 12$.

Example C   Find $\dfrac{52x^2 - 64}{4}$.

The details are

$$
\begin{array}{r}
13x^2 - 16 \\
4\overline{\smash{\big)}52x^2 - 64} \\
\underline{52x^2} \\
-64 \\
\underline{-64} \\
0
\end{array}
$$

Check: $4(13x^2 - 16) = 52x^2 - 64$. Thus, $\dfrac{52x^2 - 64}{4} = 13x^2 - 16$.

Example D   Find $\dfrac{35x^2 - 56x + 14}{7}$.

$$
\begin{array}{r}
5x^2 - 8x + 2 \\
7\overline{\smash{\big)}35x^2 - 56x + 14} \\
\underline{35x^2} \\
-56x \\
\underline{-56x} \\
14 \\
\underline{14} \\
0
\end{array}
$$

The next examples have monomial divisors with variables.

Example E   Find $\dfrac{96m^2 + 32m}{8m}$.

$$
\begin{array}{r}
12m + 4 \\
8m\overline{\smash{\big)}96m^2 + 32m} \\
\underline{96m^2} \\
32m \\
\underline{32m} \\
0
\end{array}
$$

Example F   Find $\dfrac{12t^6 - 36t^4 - 3t^2}{3t^2}$.

$$
\begin{array}{r}
4t^4 - 12t^2 - 1 \\
3t^2\overline{\smash{\big)}12t^6 - 36t^4 - 3t^2} \\
\underline{12t^6} \\
-36t^4 \\
\underline{-36t^4} \\
-3t^2 \\
\underline{-3t^2} \\
0
\end{array}
$$

B1.7 PROBLEMS

Divide to complete each equation. Show your work in the space provided.

1. $\dfrac{12x + 9}{3} = $ _4x+3_

$4x+3$

2. $\dfrac{15y - 10}{5} = $ _____

3. $\dfrac{20t - 16}{4} = $ _5t+4_

4. $\dfrac{10s + 14}{2} = $ _____

5. $\dfrac{91m^2 - 21m}{7} = $ _13m²-3m_

6. $\dfrac{18k^2 - 108k}{9} = $ _____

7. $\dfrac{-336p^4 + 48p^3 - 16}{8} = $ _-42p⁴+6p³-2_

8. $\dfrac{-414q^5 - 36q^2 + 42}{6} = $ _____

9. $\dfrac{84s^3 - 78s^2}{3s} = $ _27s²-26s_

10. $\dfrac{26w^2 + 91w}{13w} = $ _____

11. $\dfrac{51z^5 + 85z^3}{17z} = $ _____

12. $\dfrac{57x^4 - 87x^2}{3x} = $ _____

13. $\dfrac{-144u^4 - 132u^3}{12u^2} = $ _____

14. $\dfrac{-225t^5 + 195t^3}{15t^3} = $ _____

15. $\dfrac{-132q^6 + 165q^4}{11q^3} =$ _____

16. $\dfrac{-154v^5 - 126v^4}{14v^3} =$ _____

17. $\dfrac{24y^4 - 18y^3 + 12y^2}{6y^2} =$ _____

18. $\dfrac{-40m^6 + 16m^5 - 16m^3}{8m^3} =$ _____

19. $\dfrac{-36n^7 - 9n^5 - 27n^3}{9n^3} =$ _____

20. $\dfrac{98p^4 - 7p^3 + 14p^2}{7p^2} =$ _____

21. $\dfrac{-102v^4 + 6v^3 - 96v^2}{6v^2} =$ _____

22. $\dfrac{-176x^8 - 132x^6 + 88x^5}{11x^4} =$ _____

23. $\dfrac{195m^7 - 120m^6 - 10m^4}{5m^4} =$ _____

24. $\dfrac{-64k^8 + 52k^6 - 32k^5}{4k^5} =$ _____

25. $\dfrac{-48t^8 - 4t^7 + 16t^6}{8t^5} =$ _____

26. $\dfrac{144u^7 - 3u^6 - 12u^5}{6u^4} =$ _____

27. $\dfrac{72x^9 - 6x^8 + 8x^7 + 12x^6}{12x^6} =$ _____

28. $\dfrac{-96y^8 + 12y^7 - 8y^6 - 32y^5}{16y^5} =$ _____

B1.8 DIVISION OF POLYNOMIALS WITH A LINEAR BINOMIAL DIVISOR

In this unit, we will no longer just have monomial divisors, but the divisors will still be simple binomials. Recall that a binomial is a polynomial with two terms. A **linear** binomial is of the form $Ax + B$, where A and B are numbers.

Example A Find $\dfrac{x^2 + 4x + 3}{x + 3}$.

Divide the x^2 term by the x term in the divisor. Put the result, in this case x, above the first term.

$$
\begin{array}{r}
x \\
x + 3 \overline{\smash{\big)}\, x^2 + 4x + 3}
\end{array}
$$

Now multiply $x + 3$ by x and subtract, as shown here:

$$
\begin{array}{r}
x \\
x + 3 \overline{\smash{\big)}\, x^2 + 4x + 3} \\
\underline{x^2 + 3x} \\
x + 3
\end{array}
$$

Finally, divide $x + 3$ into the $x + 3$ that resulted. Thus, put $+1$ above the dividend, multiply, and subtract.

$$
\begin{array}{r}
x \ + 1 \\
x + 3 \overline{\smash{\big)}\, x^2 + 4x + 3} \\
\underline{x^2 + 3x} \\
x + 3 \\
\underline{x + 3} \\
0
\end{array}
$$

Check by multiplying: $(x + 3)(x + 1) = x^2 + 4x + 3$. Thus,

$$
\frac{x^2 + 4x + 3}{x + 3} = x + 1
$$

Example B Find $\dfrac{2x^2 - x - 6}{x - 2}$.

Follow the same procedure. Check to make sure that you "see" each step.

$$
\begin{array}{r}
2x \ + 3 \\
x - 2 \overline{\smash{\big)}\, 2x^2 - \ x - 6} \\
\underline{2x^2 - 4x} \\
3x - 6 \\
\underline{3x - 6} \\
0
\end{array}
$$

Check by multiplying: $(x - 2)(2x + 3) = 2x^2 - x - 6$.

131

Example C Find $\dfrac{3x^3 + 2x + 5}{x + 1}$.

We set the division up in almost the usual way. However, notice that there is no x^2 term in the numerator. But the process itself may lead to an x^2 term. Thus, we'll provide for the possibility by inserting $0 \cdot x^2$ as shown.

$$
\begin{array}{r}
3x^2 - 3x\ + 5 \\
x + 1\overline{\smash{\big)}\,3x^3 + 0x^2 + 2x + 5} \\
\underline{3x^3 + 3x^2 } \\
-3x^2 + 2x \\
\underline{-3x^2 - 3x } \\
5x + 5 \\
\underline{5x + 5} \\
0
\end{array}
$$

Check:

$$(x + 1)(3x^2 - 3x + 5) = 3x^2(x + 1) - 3x(x + 1) + 5(x + 1)$$
$$= 3x^3 + 3x^2 - 3x^2 - 3x + 5x + 5$$
$$= 3x^3 + 2x + 5$$

Example D Find $\dfrac{4x^3 - 5x^2 - 12}{x - 2}$.

This time, there is no x term in the numerator. Insert $0 \cdot x$, and proceed as before.

$$
\begin{array}{r}
4x^2 + 3x\ + 6 \\
x - 2\overline{\smash{\big)}\,4x^3 - 5x^2 + 0x - 12} \\
\underline{4x^3 - 8x^2 } \\
3x^2 + 0x \\
\underline{3x^2 - 6x } \\
6x - 12 \\
\underline{6x - 12} \\
0
\end{array}
$$

Example E Find $\dfrac{16x^4 - 81}{2x + 3}$.

This time, we insert several terms with 0 coefficients.

$$
\begin{array}{r}
8x^3 - 12x^2 + 18x\ - 27 \\
2x + 3\overline{\smash{\big)}\,16x^4 + \ 0x^3 + \ 0x^2 + \ 0x - 81} \\
\underline{16x^4 + 24x^3 } \\
-24x^3 + \ 0x^2 \\
\underline{-24x^3 - 36x^2 } \\
36x^2 + \ 0x \\
\underline{36x^2 + 54x } \\
-54x - 81 \\
\underline{-54x - 81} \\
0
\end{array}
$$

B1.8 PROBLEMS

Divide to complete each equation. Show your work.

1. $\dfrac{x^2 + 3x + 2}{x + 2} =$ _____

2. $\dfrac{2x^2 - 3x + 1}{x - 1} =$ _____

3. $\dfrac{3x^2 - 4x + 1}{x - 1} =$ _____

4. $\dfrac{x^2 + 6x + 5}{x + 5} =$ _____

5. $\dfrac{2x^2 + x - 1}{2x - 1} =$ _____

6. $\dfrac{3x^2 + 5x - 2}{3x - 1} =$ _____

7. $\dfrac{3x^2 - 7x - 6}{3x + 2} =$ _____

8. $\dfrac{6x^2 + 7x - 3}{2x + 3} =$ _____

9. $\dfrac{3x^3 - 10x - 4}{x - 2} =$ _____

10. $\dfrac{2x^3 - 15x - 9}{x - 3} =$ _____

133

11. $\dfrac{4x^4 - 5x^3 - 8x + 10}{4x - 5} =$ _____

12. $\dfrac{6x^4 - 14x^3 - 9x + 21}{3x - 7} =$ _____

13. $\dfrac{81x^4 - 1}{3x - 1} =$ _____

14. $\dfrac{16x^4 - 1}{2x - 1} =$ _____

15. $\dfrac{x^4 - 4x^3 + 6x^2 - 4x + 1}{x - 1} =$ _____

16. $\dfrac{x^3 - 3x^2 + 3x - 1}{x - 1} =$ _____

17. $\dfrac{x^3 + 3x^2 + 3x + 1}{x + 1} =$ _____

18. $\dfrac{x^4 + 4x^3 + 6x^2 + 4x + 1}{x + 1} =$ _____

19. $\dfrac{x^5 - 1}{x - 1} =$ _____

20. $\dfrac{x^3 + 1}{x + 1} =$ _____

21. $\dfrac{x^4 - 1}{x - 1} =$ _____

22. $\dfrac{x^5 + 1}{x + 1} =$ _____

B1.9 FACTORING POLYNOMIALS BY FINDING COMMON FACTORS

When two or more numbers are multiplied, the answer is their **product**. The numbers you begin with are called **factors**. When beginning with a single number, you can ask, "Can this be written as a product of factors?" If it can, then you are said to be **factoring** the number.

Example A   Factor 12 in all possible ways as a product of two positive whole numbers.

Here's the list:

$$1 \cdot 12 \qquad 4 \cdot 3$$
$$2 \cdot 6 \qquad 6 \cdot 2$$
$$3 \cdot 4 \qquad 12 \cdot 1$$

The product $1 \cdot 12$ is called the **trivial factorization**, since every number is 1 times itself. Also, the pair $1 \cdot 12$ and $12 \cdot 1$, the pair $2 \cdot 6$ and $6 \cdot 2$, and the pair $3 \cdot 4$ and $4 \cdot 3$ are considered to be the same. Thus, our list reduces to three:

$$1 \cdot 12 \qquad 2 \cdot 6 \qquad 3 \cdot 4$$

These remarks, ideas, and terminology also hold for polynomials.

Example B   Factor the monomial $6x^2$ in all possible ways as a product of two monomials using only positive whole numbers and variables.

Read all the restrictions before checking the list. We begin the list with the trivial factorization.

$$1 \cdot 6x^2 \qquad 2 \cdot 3x^2$$
$$6 \cdot x^2 \qquad 3 \cdot 2x^2$$
$$6x \cdot x \qquad 2x \cdot 3x$$

The purpose of this and the next two units is to factor polynomials with two or more terms. In this unit, **common factors**—that is, factors that are common (the same) to each term of the polynomial—will be covered.

Example C   Factor $4x^2 + 8x$.

This factors in many ways. Here are some:

$$2(2x^2 + 4x) \qquad\qquad 2x(2x + 4)$$
$$x(4x + 8) \qquad\qquad 4x(x + 2)$$

Notice that $1(4x^2 + 8x)$ was omitted; however, soon you'll see that we do have a use for the trivial factorization. Of the four factorizations in the above list, the last one is *the* answer. This is because all the others have the property that the terms of the binomial have other factors in common.

135

Example D   Factor $25st^2 + 5s$.

The factor $5s$ is common to each term. Thus,

$$25st^2 = 5s(5t^2 + 1)$$

Notice that we used here the trivial factorization of the second term.

Example E   Factor $8x^3y - 12x^2y^2 + 24xy^2$.

You get common factors "by inspection" of the terms. Suppose here that you factor out $2x$. Then you would write the product

$$2x(4x^2y - 6xy^2 + 12y^2)$$

But you now notice that the terms in the parentheses still have $2y$ in common. Thus, you correct your first effort and write

$$4xy(2x^2 - 3xy + 6y)$$

Example F   Factor $9a^2(a - 2) + 4a - 8$.

At first, this looks impossible. But since $4a - 8 = 4(a - 2)$, we can rewrite it:

$$9a^2(a - 2) + 4(a - 2)$$

Now $(a - 2)$ is a common factor. Thus, $(a - 2)(9a^2 + 4)$ is the factorization.

Example G   Factor $3mn - n - 6m^2 + 2m$.

The first two terms have a common monomial factor as do the third and fourth terms.

$$3mn - n - 6m^2 + 2m = n(3m - 1) - 2m(3m - 1)$$
$$= (3m - 1)(n - 2m)$$

B1.9 PROBLEMS

Use the methods of this unit to factor each polynomial.

1. $5x - 10 =$ $\underline{5(x-2)}$

2. $12m - 8 =$ $\underline{3(m-2)}$

3. $6 + 6p^2 =$ $\underline{6(1+p^3)}$

4. $3a + 3b =$ _____

5. $2r - 2st =$ $\underline{2(r-st)}$

6. $4x^2 + 4 =$ _____

7. $9a^2 + 48ab =$ $\underline{3a(3a+16b)}$

8. $34p^2 - 51pq =$ _____

9. $57cd - 38d^2 =$ $\underline{19d(3c-2d)}$

10. $27rs + 42s^2 =$ _____

11. $3m^3 + m^2 + 3m =$ $\underline{m(3m^2+m+3)}$

12. $6ab^2 + 9a^2b - 12ab =$ _____

13. $12p^3q - 18p^2q + 24pq^3 =$ $\underline{6pq(p^2-3p+4q^2)}$

14. $5x^4 + x^3 + 4x^2 =$ _____

15. $2c^5 + 4c^4 + 6c^3 + 8c^2 =$ $\underline{2c^2(c^3+2c^2+3c+4)}$

16. $3a + 6a^2 + 9a^3 + 12a^4 =$ _____

17. $4rs + 6r^2 - 8rs^2 =$ $\underline{2rs(2s+3r-4s)}$

18. $9u^3v - 6u^2v^2 + 15uv^2 =$ _____

137

19. $98a^5b^3c^4 - 42a^3b^6c^2 - 63a^4b^4c^3 =$ _____

20. $-52x^7y^5z^6 + 91x^4y^6z^3 - 65x^5y^4z^4 =$ _____

21. $x(u - v) + y(u - v) =$ _____

22. $p(s + t) - q(s + t) =$ _____

23. $a(2 + r) + 4 + 2r =$ _____

24. $2b(s - 3) + 3s - 9 =$ _____

25. $c(2d - 3) - 4(3 - 2d) =$ _____

26. $5(m - 6n) + x(6n - m) =$ _____

27. $ax - ay + bx - by =$ _____

28. $us + vs + ut + vt =$ _____

29. $ab^2 + b + ab + 1 =$ _____

30. $cd - 1 + cd^2 - d =$ _____

31. $m^2 - 1 + 5m - 5 =$ _____

32. $m^2 + 6mn + 3m + 18n =$ _____

33. $4c + d + 16c^2 + 4cd =$ _____

34. $3x - 6 + 9x^2 - 18x =$ _____

35. $s^2t^2 + s^2 - 3t^2 - 3 =$ _____

36. $18 + 6m - 3n^2 - mn^2 =$ _____

37. $8a^2bc - 10ac - 12ab^2 + 15b =$ _____

38. $18x^3 - 15axy - 12x^2y + 10ay^2 =$ _____

B1.10 FACTORING A TRINOMIAL

As you will see, a given trinomial might be: (1) Factorable as the square of a binomial; (2) Factorable as the product of unlike binomials; or (3) Unable to be factored at the level of this book.

Square of a Binomial

Recall that $(A + B)^2 = A^2 + 2AB + B^2$

A trinomial is factorable as the square of a binomial only if it fits this pattern.

Example A   Factor $16x^2 + 24x + 9$.

Note that

$$16x^2 = (4x)^2 \qquad 24x = (2)(4x)(3) \qquad 9 = 3^2$$

Thus, $16x^2 + 24x + 9 = (4x)^2 + (2)(4x)(3) + 3^2$.

It fits the pattern. Consequently, $16x^2 + 24x + 9 = (4x + 3)^2$.

Example B   Is $6x + x^2 + 9$ factorable as the square of a binomial?

Rewrite it as $x^2 + 6x + 9$. It fits the pattern, and factors as $(x + 3)^2$.

Example C   Is $4x^2 - 17x + 4$ factorable as the square of a trinomial?

The first and last terms are squares, but the middle term does not fit the pattern. The answer is no. However, this does factor, as you will see.

Product of Unlike Binomials

Actually, there is a pattern for Example C, but we will use trial-and-error.

Example D   Factor $4x^2 - 17x + 4$.

First, factor $4x^2$, listing all the ways to get two binomials each with a factor of x.

$$(4x \quad)(x \quad) \quad \text{and} \quad (2x \quad)(2x \quad)$$

Now factor the third term, 4. Combine each possibility with each of the above in each possible way.

$$(4x - 4)(x - 1) \qquad (2x - 4)(2x - 1)$$

$$(4x - 1)(x - 4) \qquad (2x - 2)(2x - 2)$$

$$(4x - 2)(x - 2)$$

Multiplication shows that the right choice is $(4x - 1)(x - 4)$ **139**

Example E   Factor $6x^2 + 13x + 5$.

First, factor $6x^2$ for the possible binomials, with each containing a factor of x.

$(x\quad)(6x\quad)$ and $(2x\quad)(3x\quad)$

Next, factor 5 (you'll need the trivial factorization).

$(x + 1)(6x + 5)\qquad(2x + 1)(3x + 5)$

$(x + 5)(6x + 1)\qquad(2x + 5)(3x + 1)$

The $+$ signs are inserted because the middle and last terms are $+$. Finally, use FOIL to get the right combination, which is $(2x + 1)(3x + 5)$.

Example F   Factor $6x^2 + 11x - 10$.

The last term is negative, so the factors look like $(\quad+\quad)$ and $(\quad-\quad)$, thus increasing the number of possibilities. We list them.

$(x\quad1)(6x\quad10)$ NO $(2x\quad1)(3x\quad10)$

$(x\quad2)(6x\quad5)$ $(2x\quad2)(3x\quad5)$ NO

$(x\quad5)(6x\quad2)$ NO $(2x\quad5)(3x\quad2)$

$(x\quad10)(6x\quad1)$ $(2x\quad10)(3x\quad1)$ NO

Leaving the $+$ and $-$ signs alone, we omit the NO cases. They lead to an even middle term, but 11 is odd. There are four cases left. Since the middle term is $+$, put the $+$ sign with the larger last term and the $-$ sign with the smaller.

$(x - 2)(6x + 5)\qquad(2x - 1)(3x + 10)$

$(x + 10)(6x - 1)\qquad(2x + 5)(3x - 2)$

A quick check shows $(2x + 5)(3x - 2)$ to be correct.

Unable to Be Factored

Example G   Show that $6x^2 + 11x + 10$ cannot be factored.

This is almost like Example E. Since all terms are $+$, the only chances are

$(x + 1)(6x + 10)\qquad(2x + 1)(3x + 10)$

$(x + 2)(6x + 5)\qquad(2x + 2)(3x + 5)$

$(x + 5)(6x + 2)\qquad(2x + 5)(3x + 2)$

$(x + 10)(6x + 1)\qquad(2x + 10)(3x + 1)$

None of them works, so $6x^2 + 11x + 10$ cannot be factored with real number coefficients.

B1.10 PROBLEMS

Factor each trinomial, where possible. If it is not factorable, write N.F. in the space provided.

1. $x^2 + 3x + 2 = \underline{(x+2)(x+1)}$ 2. $x^2 + 4x + 3 = \underline{\hspace{2cm}}$

3. $x^2 + 2x + 1 = \underline{(x+1)^2}$ 4. $x^2 + 4x + 4 = \underline{\hspace{2cm}}$

5. $9x^2 + 6x + 1 = \underline{(3x+1)^2}$ 6. $x^2 + 3x + 3 = \underline{\hspace{2cm}}$

7. $2x^2 + 2x + 1 = \underline{\hspace{2cm}}$ 8. $16x^2 + 8x + 1 = \underline{\hspace{2cm}}$

9. $x^2 - 5x + 6 = \underline{\hspace{2cm}}$ 10. $x^2 - 10x + 25 = \underline{\hspace{2cm}}$

11. $6x^2 - 3x + 4 = \underline{\hspace{2cm}}$ 12. $x^2 - 2x - 35 = \underline{\hspace{2cm}}$

13. $2x^2 + 7x - 15 = \underline{\hspace{2cm}}$ 14. $x^2 - 6x + 8 = \underline{\hspace{2cm}}$

15. $x^2 - 12x + 36 = \underline{\hspace{2cm}}$ 16. $25x^2 + 36 - 60x = \underline{\hspace{2cm}}$

17. $x^2 - 11x - 42 = \underline{\hspace{2cm}}$ 18. $4x^2 - 5x + 6 = \underline{\hspace{2cm}}$

19. $49 + 16x^2 - 56x = \underline{\hspace{2cm}}$ 20. $2x^2 + x - 10 = \underline{\hspace{2cm}}$

141

21. $3x^2 - 5x - 2 =$ _____

22. $x^2 + x + 1 =$ _____

23. $5x^2 + 10x + 5 =$ _____
(*Hint*: Factor out a 5 first)

24. $16x^2 + 22x - 3 =$ _____

25. $6x^2 - 7x + 2 =$ _____

26. $4x^2 - 5x + 3 =$ _____

27. $x^2 - x - 1 =$ _____

28. $3x^2 - 8x - 3 =$ _____

29. $21x^2 + x - 2 =$ _____

30. $7x^2 - 14x + 7 =$ _____
(*Hint*: Factor out a 7 first)

31. $4x^2 - 7x + 4 =$ _____

32. $2x^2 + 7x + 3 =$ _____

33. $2x^2 - 3x + 5 =$ _____

34. $2x^2 - x - 21 =$ _____

35. $12x^2 + 11x - 15 =$ _____

36. $12x^2 - 7x + 12 =$ _____

37. $16x^2 - 16x - 12 =$ _____

38. $12x^2 - 11x - 15 =$ _____

39. $9x^2 + 15x - 14 =$ _____

40. $3x^2 - 4x + 7 =$ _____

B1.11 FACTORING A DIFFERENCE OF SQUARES

As the title states, here we will be factoring the difference of two squares, such as $x^2 - y^2$, $4a^2 - 16b^2$, and $1 - 9x^2$. Before beginning, make sure that the two terms are indeed squares and that they are separated by a minus sign. If not both of these conditions are met, then the expression cannot be factored as a difference of squares. If the conditions are met, then the process of finding factors is quite simple.

Example A Factor $x^2 - y^2$.

This factors as $(x + y)(x - y)$. The FOIL process shows that this is correct:

$$(x + y)(x - y) = x^2 - xy + xy - y^2$$
$$\uparrow \qquad \uparrow \qquad \uparrow \qquad \uparrow$$
$$\text{F} \qquad \text{O} \qquad \text{I} \qquad \text{L}$$

$$= x^2 - y^2$$

Example B Factor $4a^2 - 9b^2$.

This factors as $(2a + 3b)(2a - 3b)$. You can use FOIL to check it.

Example C Factor $1 - 16x^2$.

$$1 - 16x^2 = (1 + 4x)(1 - 4x)$$

The square terms can take a variety of forms. The key is to insure that they are squares. Here is an example using fractions.

Example D $\dfrac{4}{9}x^2 - \dfrac{36}{49}y^2 = \left(\dfrac{2}{3}x - \dfrac{6}{7}y\right)\left(\dfrac{2}{3}x + \dfrac{6}{7}y\right).$

The following example involves the use of exponents in coefficients of the x and y terms.

Example E $a^2x^2 - b^2y^2 = (ax - by)(ax + by).$

It is now possible to generalize this process of factoring the differences of two squares to include many different forms. Here are some additional examples that you check by multiplying.

Example F $x^4 - y^4 = (x^2 - y^2)(x^2 + y^2).$

In this example, notice that you can factor further since $x^2 - y^2$ is a difference of squares. Thus, $x^4 - y^4 = (x - y)(x + y)(x^2 + y^2).$

Example G $64a^6 - b^4 = (8a^3 - b^2)(8a^3 + b^2).$

143

Example H $25s^2t^8 - 36n^6 = (5st^4 - 6n^3)(5st^4 + 6n^3)$.

There are expressions that on the surface do not appear to be factorable by finding differences of squares, but some of these may contain factors that can be factored out before the perfect squares are evident.

Example I   Factor $8x^2 - 18$.

Neither 8 nor 18 is a square, but both contain the factor 2.

$$8x^2 - 18 = 2(4x^2 - 9) = 2(2x - 3)(2x + 3)$$

Knowing about factoring differences of squares can lead to some tricks for mental numerical calculation.

Example J   What is the product $49 \cdot 51$?

Rewrite this as $(50 - 1)(50 + 1)$. Then you know that

$$49 \cdot 51 = (50 - 1)(5 + 1)$$

$$= 50^2 - 1^2 = 2500 - 1 = 2499$$

You can surely do this in your head.

Example K   What is $33 \cdot 27$?

$$33 \cdot 27 = (30 + 3)(30 - 3)$$

$$= 30^2 - 3^2 = 900 - 9 = 891$$

Example L   In a geometry problem involving the Pythagorean theorem, one wants to know the value of $2641^2 - 2640^2$. Calculate this quickly.

$$2641^2 - 2640^2 = (2641 - 2640)(2641 + 2640)$$

$$= 1 \cdot (2641 + 2640)   = 5281$$

This is a lot faster than squaring and subtracting.

Example M   Find $103^2 - 100^2$.

$$103^2 - 100^2 = (103 - 100)(103 + 100)   = 3 \cdot 203   = 609$$

B1.11 PROBLEMS

The following are partially factored. Fill in the blanks.

1. $a^2 - b^2 = (a - \_)(a + \_)$

2. $x^2 - y^2 = (\_ - y)(x + \_)$

3. $4a^2 - 9y^2 = (\_a - 3y)(\_a + \_y)$

4. $16b^2 - 25h^2 = (4\_ - \_h)(\_ + 5h)$

5. $81 - 36z^2 = (\_ - \_z)(\_ + \_z)$

6. $144x^2 - 1 = (\_x - \_)(\_x + \_)$

7. $9y^2 - 9x^2 = (3\_ - 3\_)(\_y + \_x)$

8. $4x^4 - 25z^4 = (2\_ - 5\_)(2\_ + \_z^2)$

9. $16a^6 - b^8 = (4\_ - \_)(4\_ + \_)$

10. $8x^2 - 18y^2 = \_(2\_ - \_)(2\_ + 3\_)$

11. $27a^2 - 3b^2 = \_(\_a - \_b)(\_a + \_b)$

12. $12x^2 - 12y^2 = \_(\_x - \_y)(\_x + \_y)$

13. $3p^2 - 12q^2 = \_(\_p - \_q)(\_p + \_q)$

14. $5x^2 - 5y^2 = \_(\_x + \_y)(\_x + \_y)$

Factor each of the following completely. If not factorable, write N.F.

15. $16x^2 - 144y^2 =$ _____

16. $64w^2 - 81z^2 =$ _____

17. $114q^2 - 81p^6 =$ _____

18. $25xp^2 - 16xr^4 =$ _____

19. $12y^3 - 192y =$ _____

20. $5p^3 - 5p =$ _____

21. $25x^2 - 16y^3 =$ _____

22. $18a^2 - 32 =$ _____

23. $8a^3b - 18ab^3 =$ _____   24. $25x^2 - 5 =$ _____

25. $a^2b^2 - 9 =$ _____   26. $x^2y^4 - x^2y^6 =$ _____

27. $16^2 - 9^2 =$ _____   28. $-x^2 + y^2 =$ _____

29. $3xa^2 - 6b^2x =$ _____   30. $(a + b)^2 - (x + y)^2 =$ _____

31. $16x^6 - 49y^9 =$ _____   32. $5x^2 - 25y^2 =$ _____

33. $-49x^2 + 64y^2 =$ _____   34. $4a^2b^2 - 9x^8 =$ _____

35. $2z^5 - 18z =$ _____   36. $a^2b^4 - p^6q^8 =$ _____

37. $pqx^2 - 4pq^3y^2 =$ _____   38. $x^2 - (p + q)^2 =$ _____

Use the methods of this unit to perform each calculation. Do it mentally if possible.

39. $39 \cdot 41 =$ _____   40. $21 \cdot 19 =$ _____

41. $101 \cdot 99 =$ _____   42. $199 \cdot 201 =$ _____

43. $56 \cdot 64 =$ _____   44. $34 \cdot 26 =$ _____

45. $60^2 - 59^2 =$ _____   46. $90^2 - 89^2 =$ _____

47. $75^2 - 70^2 =$ _____   48. $105^2 - 100^2 =$ _____

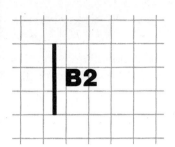

B2

OPERATIONS WITH RATIONAL EXPRESSIONS

B2.1 SIMPLIFICATION OF A RATIONAL EXPRESSION BY CANCELLATION OF COMMON FACTORS

One of the first concepts learned about fractions is how to simplify them, meaning to write them so that the numerator and denominator have no common factors other than 1. Keep in mind that a fraction is also called a rational number. The key to remember is that the number 1, when written as a fraction, can be written in many ways: $\frac{1}{1}$, $\frac{3}{3}$, $\frac{23}{23}$, and so on. Here are examples of simplifying rational expressions.

Example A $\frac{4}{8} = \frac{4 \cdot 1}{4 \cdot 2} = \frac{4}{4} \cdot \frac{1}{2} = \frac{1}{2}$.

$\frac{4}{4}$ is one way of writing 1 as a fraction.

In algebra, we frequently need to simplify rational expressions that are basically fractions, but that contain both numbers and literal terms. This means that we will need to make use of the skills acquired for factoring algebraic expressions. Again, recall that 1 can be written in many ways, including the use of letters: $\frac{a}{a}$, $\frac{xy}{xy}$, $\frac{pqt}{pqt}$, $\frac{2z}{2z}$, etc. However, the letters cannot represent 0 because 0 as a denominator yields a meaningless expression.

Example B $\frac{4a}{6a^2} = \frac{2 \cdot 2 \cdot a}{2 \cdot 3 \cdot a \cdot a} = \frac{2}{3a}$.

This is true because $\frac{2}{2}$ equals 1, and $\frac{a}{a}$ equals 1. This leaves $\frac{2}{3a}$.

151

The process of factoring to find pairs of numerators and denominators that are equal to 1 is called simplifying rational expressions. It is also called cancellation of common factors.

Example C $\quad \dfrac{6x}{9xy} = \dfrac{2}{3y}$.

$\dfrac{6}{9}$ simplifies to $\dfrac{2}{3}$ and $\dfrac{x}{x}$ is another way of writing 1.

Example D $\quad \dfrac{12x^3y^5}{30x^2y^4z} = \dfrac{(6)(2)(x^2)(x)(y^4)(y)}{(6)(5)(x^2)(y^4)(z)}$

$$= \dfrac{2xy}{5z}$$

Notice it is possible to cancel when parts of numerators and denominators form a rational expression (fraction) that equals 1.

Now consider what happens when the numerators and denominators contain polynomials. Again, we begin by factoring as completely as possible before cancelling.

Example E $\quad \dfrac{2a^3}{6a + 8b} = \dfrac{2a^3}{2(3a + 4b)}$

$$= \dfrac{a^3}{3a + 4b}$$

The above example illustrates the care that must be taken when factoring. Factors to be cancelled must be those that are multiplied. A very common error is to cancel incorrectly in this example, giving a wrong answer of $\dfrac{2a^2}{6 + 8b}$.

Example F $\quad \dfrac{x^2 + 2xy + y^2}{x^2 + xy} = \dfrac{(x + y)(x + y)}{x(x + y)}$

$$= \dfrac{x + y}{x}$$

Here the cancelled factor is $(x + y)$.

Example G $\quad \dfrac{4x^2 - 9y^2}{10xy + 15y^2} = \dfrac{(2x - 3y)(2x + 3y)}{5y(2x + 3y)}$

$$= \dfrac{2x - 3y}{5y}$$

There are times when you may have to make sign changes in the rational expression in order to cancel.

Example H

$$\frac{2 - x}{x^2 - 4} = -\frac{(-2 + x)}{(x - 2)(x + 2)}$$

$$= \frac{-(x - 2)}{(x - 2)(x + 2)}$$

$$= -\frac{1}{(x + 2)}$$

$$= \frac{-1}{x + 2}$$

Example I

$$\frac{(4a + 4b)(a - b)}{b - a} = \frac{4(a + b)(a - b)}{b - a}$$

$$= -\frac{-4(a + b)(a - b)}{-(b - a)}$$

$$= -\frac{-4(a + b)(a - b)}{a - b}$$

$$= -4(a + b)$$

In the last two examples, you see that there are several forms for the final answer, but normally the negative sign will be placed in the numerator. The final word of caution is: Do not begin to cancel until the numerator and denominator have been factored completely.

Example J

$$\frac{4x^2 - 4xy + y^2}{6x^2 - 5xy + y^2} = \frac{(2x - y)(2x - y)}{(2x - y)(3x - y)}$$

$$= \frac{2x - y}{3x - y}$$

Example K

$$\frac{x^4 - y^4}{y^2 - x^2} = \frac{(x^2 - y^2)(x^2 + y^2)}{(y - x)(y + x)}$$

$$= \frac{(x - y)(x + y)(x^2 + y^2)}{(y - x)(y + x)}$$

$$= -x^2 - y^2$$

Make sure that you followed the signs in Example K.

Example L $$\frac{6r^2 - 5rs - 6s^2}{9s^2 - 12rs + 4r^2} = \frac{(2r - 3s)(3r + 2s)}{(3s - 2r)^2}$$

$$= \frac{(2r - 3s)(3r + 2s)}{(2r - 3s)^2}$$

$$= \frac{(2r - 3s)(3r + 2s)}{(2r - 3s)(2r - 3s)}$$

$$= \frac{3r + 2s}{2r - 3s}$$

In Example L, make sure you see why $(3s - 2r)^2 = (2r - 3s)^2$. If you don't, just multiply them out.

Example M $$\frac{8a^2 - 6ab + b^2}{b^2 - ab - 2a^2} = \frac{(4a - b)(2a - b)}{(b - 2a)(b + a)}$$

$$= \frac{b - 4a}{b + a}$$

The problems are easier than some of these examples, but these examples will be useful later on. Be sure to study them.

B2.1 PROBLEMS

Simplify.

1. $\dfrac{15}{30} = $ _____

2. $\dfrac{27}{81} = $ _____

3. $\dfrac{-2x}{12x} = $ _____

4. $\dfrac{-5w}{15wz} = $ _____

5. $\dfrac{6ab}{8a} = $ _____

6. $\dfrac{-14}{28x} = $ _____

7. $\dfrac{24t}{36} = $ _____

8. $\dfrac{-ab}{-12a} = $ _____

9. $\dfrac{4xy}{4xy} = $ _____

10. $\dfrac{12p^2}{-18p} = $ _____

11. $\dfrac{22ab}{-33b} = $ _____

12. $\dfrac{45xy}{60xz} = $ _____

13. $\dfrac{-8ab^2}{10ab} = $ _____

14. $\dfrac{21ux^2}{28x} = $ _____

15. $\dfrac{-56x^3}{-64x} = $ _____

16. $\dfrac{100p}{1000p^2} = $ _____

17. $\dfrac{-75q^3}{25q} = $ _____

18. $\dfrac{10p^2q^2}{-45pq^2} = $ _____

19. $\dfrac{28x}{35xy} = $ _____

20. $\dfrac{-q^3}{-q^2} = $ _____

21. $\dfrac{a+b}{a^2+2ab+b^2} = $ _____

22. $\dfrac{x^2-y^2}{(x+y)^2} = $ _____

23. $\dfrac{3x-6}{4x-8} = $ _____

24. $\dfrac{3m^3}{6m+3m} = $ _____

25. $\dfrac{p^2-6p+8}{p-2} = $ _____

26. $\dfrac{3q-6}{q^2-4} = $ _____

27. $\dfrac{a-b}{b-a} = $ _____

28. $\dfrac{w^2-z^2}{z-w} = $ _____

29. $\dfrac{x^2-y^2}{y^2-2xy+x^2} = $ _____

30. $\dfrac{a^2+a-6}{2-a} = $ _____

B2.2 EVALUATION OF A RATIONAL EXPRESSION

Algebra makes use both of numbers and of letters that represent numbers. It is the ability to use letters for numbers that gives algebra its power. Expressions that contain letters, such as $a + b$, xy, $p - q$, and $\dfrac{z}{w}$ will each have numerical values when numbers are substituted for the letters. Thus, $a + b$ can represent any of infinitely many values, depending upon what a is and what b is.

We will now evaluate rational expressions when numerical values are given to the various letters. Doing this properly depends upon the correct use of *order of operations*. These are exactly the same as those used in arithmetic, which are:

1. If there are exponents, these should be done first. If the exponents apply to parentheses, then clear the parentheses first.

2. Next, do the multiplication and division from left to right.

3. Then do the addition and subtraction from left to right. If parentheses are involved:

4. First do all the operations inside the parentheses according to steps 1 and 2.

5. Then do all the remaining operations according to the above order.

Example A Evaluate $a + bc$ when $a = 4$, $b = 5$, $c = 6$.

$4 + 5 \cdot 6$ Multiply $5 \cdot 6 = 30$
       Add $4 + 30 = 34$

Example B Evaluate $\dfrac{x}{y} + 2z - y$ when $x = 4$, $y = 2$, $z = 6$.

$\dfrac{4}{2} + 2 \cdot 6 - 2$ Divide $\dfrac{4}{2} = 2$
$2 + 2 \cdot 6 - 2$ Multiply $2 \cdot 6 = 12$
$2 + 12 - 2$ Add $2 + 12 = 14$
$14 - 2$ Subtract $14 - 2 = 12$
12

If it is a rational expression that you are evaluating, do everything above and below the division bar first according to the preceding rules. Also keep in mind that you may be able to simplify an expression before making substitutions.

Example C Evaluate $\dfrac{x^2 y}{xy^2}$ when $x = 4$, $y = -2$ without simplifying first.

$\dfrac{4^2(-2)}{4(-2)^2}$ $x^2 = 4^2 = 16$ $16(-2) = -32$

$y^2 = (-2)^2 = 4$ $4 \cdot 4 = 16$

$\dfrac{-32}{16} = -2$

157

Example D Evaluate $\dfrac{x^2y}{xy^2}$ when $x = 4$, $y = -2$, but simplify first, then evaluate.

$$\frac{x^2y}{xy^2} = \frac{x}{y}$$

$$\frac{4}{-2} = -2$$

You can see that a lot of work was saved by simplifying first.

Example E Evaluate $\dfrac{ab}{a - ab}$ when $a = 4$, $b = 5$.

$$\text{Simplify } \frac{ab}{a(1 - b)} = \frac{b}{1 - b}$$

$$= \frac{5}{1 - 5} = \frac{5}{-4} = -\frac{5}{4}$$

Example F Evaluate $\dfrac{a^2 - b^2}{(a - b)^2}$ when $a = 2$, $b = 3$.

$$\text{Simplify } \frac{a^2 - b^2}{(a - b)^2} = \frac{(a - b)(a + b)}{(a - b)(a - b)} = \frac{a + b}{a - b}$$

$$= \frac{2 + 3}{2 - 3} = \frac{5}{-1} = -5$$

As you have seen, there are often several approaches to evaluating a rational expression. If the expression can be simplified, then this is the desired approach. However, if the expression looks too complicated, then you can simply substitute the numerical values for the letters and compute, using the order of operations rules.

B2.2 PROBLEMS

Evaluate. Consider simplifying first.

1. $a + ab$ when $a = 5$, $b = 8$ _____

2. $xy - y$ when $x = 2$, $y = 4$ _____

3. $a(a - b)$ when $a = 9$, $b = -3$ _____

4. $pq - (p - q)$ when $p = 2$, $q = 7$ _____

5. $(a - b)(a + b)$ when $a = 5$, $b = 4$ _____

6. $p^2 - q$ when $p = 3$, $q = 9$ _____

7. $\dfrac{x}{y} + x$ when $x = 8$, $y = 4$ _____

8. $\dfrac{ab^2}{b} - ab$ when $a = 3$, $b = 5$ _____

9. $\dfrac{5t^2}{s} + ts - t$ when $s = 4$, $t = 2$ _____

10. $x - (x^2 + y)^2$ when $x = 2$, $y = 4$ _____

11. $a(b - a) - \dfrac{a}{b}$ when $a = 6$, $b = 3$ _____

12. $\dfrac{x}{y} - \dfrac{y}{x}$ when $x = 4$, $y = 2$ _____

13. $\dfrac{ab}{ab - b}$ when $a = 4$, $b = 6$ _____

14. $\dfrac{x^2 - y^2}{x + y}$ when $x = 3$, $y = 5$ _____

15. $\dfrac{a^2 - 2ab + b^2}{a(a - b)}$ when $a = 2$, $b = 3$ _____

16. $\dfrac{(p - q)^2}{p - q}$ when $p = 1$, $q = 2$ _____

17. $\dfrac{11x^2 + 66x}{11x}$ when $x = 2$ _____

18. $\dfrac{a^2 - 2ab + b^2}{a^2 + 2ab + b^2}$ when $a = 3$, $b = 4$ _____

19. $\dfrac{(t - r)^2}{r(t - r)^2}$ when $r = 8$, $t = 6$ _____

20. $\dfrac{x(xy^2)}{y^2}$ when $x = 5$, $y = 6$ _____

B2.3 ADDITION AND SUBTRACTION
OF RATIONAL EXPRESSIONS

Adding and subtracting rational expressions is similar to adding and subtracting fractions. If the denominators are the same, it is simply a matter of operating on the numerators with the common denominator.

Example A
$$\frac{1}{5} + \frac{2}{5} = \frac{1+2}{5} = \frac{3}{5} \qquad \frac{5}{7} - \frac{4}{7} = \frac{5-4}{7} = \frac{1}{7}$$

Example B
$$\frac{a}{x} + \frac{b}{x} = \frac{a+b}{x} \qquad \frac{p}{r} - \frac{q}{r} = \frac{p-q}{r}$$

When rational expressions have different denominators, they must be rewritten so the denominators are the same *before* either addition or subtraction can be performed.

Example C Add: $\dfrac{2}{a} + \dfrac{3}{b}$.

A common denominator can be ab. Thus, we will have to multiply $\dfrac{2}{a}$ by $\dfrac{b}{b}$, and $\dfrac{3}{b}$ by $\dfrac{a}{a}$.

$$\frac{2b}{ab} + \frac{3a}{ba} = \frac{2b + 3a}{ab}$$

Example D Subtract: $\dfrac{2}{x} - y$.

Multiply $\dfrac{y}{1}$ by $\dfrac{x}{x}$.

$$\frac{y}{1} \cdot \frac{x}{x} = \frac{2}{x} - \frac{yx}{x} = \frac{2 - xy}{x}$$

Example E
$$\frac{4}{a+1} + \frac{3}{a-1} = \frac{4(a-1)}{(a+1)(a-1)} + \frac{3(a+1)}{(a-1)(a+1)}$$
$$= \frac{4a - 4 + 3a + 3}{(a+1)(a-1)} = \frac{7a - 1}{a^2 - 1}.$$

At times, you may find multiplying by (-1) helpful in creating a common denominator.

Example F   Subtract: $\dfrac{5}{a-b} - \dfrac{4}{b-a}$.

Here you can multiply numerator and denominator of the second rational expression by -1.

$$\frac{5}{a-b} - \frac{(-1)4}{(-1)(b-a)} = \frac{5}{a-b} + \frac{4}{a-b} = \frac{9}{a-b}$$

In the above example, multiplying by $\dfrac{(-1)}{(-1)}$ changed $(b-a)$ to $(a-b)$ and changed the problem from subtraction to addition.

Example G   Add: $\dfrac{x-y}{xy} + \dfrac{y-z}{yz}$.

Here the common denominator is xyz.

$$\frac{z(x-y)}{xyz} + \frac{x(y-z)}{xyz} = \frac{xz - yz + xy - xz}{xyz}$$

$$= \frac{xy - yz}{xyz} = \frac{y(x-z)}{xyz}$$

$$= \frac{x-z}{xz}$$

Example H   Subtract: $\dfrac{5}{a^2b} - \dfrac{3}{ab^2}$.

Here the common denominator is a^2b^2.

$$\frac{5}{a^2b} - \frac{3}{ab^2} = \frac{5b}{a^2b^2} - \frac{3a}{a^2b^2} = \frac{5b - 3a}{a^2b^2}$$

Example I   Add: $\dfrac{2}{a^2 - 2ab + b^2} + \dfrac{2}{a-b}$.

In this example, it is possible to factor the first denominator.

$$\frac{2}{(a-b)^2} + \frac{2}{a-b} = \frac{2}{(a-b)^2} + \frac{2(a-b)}{(a-b)^2}$$

$$= \frac{2 + 2(a-b)}{(a-b)^2} = \frac{2a - 2b + 2}{(a-b)^2}$$

$$= \frac{2(a-b+1)}{(a-b)^2}$$

Adding and subtracting rational expressions may take several more steps when compared to adding and subtracting fractions. Also, there are often different forms in which the final answer can be written.

B2.3 PROBLEMS

Perform the indicated operations.

1. $\dfrac{1}{b} + \dfrac{4}{b} = $ _____

2. $\dfrac{5w}{x} + \dfrac{6w}{x} = $ _____

3. $\dfrac{x}{y} - \dfrac{4x}{y} = $ _____

4. $\dfrac{3a}{x} + \dfrac{a}{2x} = $ _____

5. $\dfrac{3x}{4y} - \dfrac{2x}{4y} = $ _____

6. $\dfrac{5}{ab} + \dfrac{8b}{a} = $ _____

7. $\dfrac{4w - 2}{w} + \dfrac{2}{w} = $ _____

8. $\dfrac{2}{3b} + \dfrac{3}{3b + 2} = $ _____

9. $\dfrac{8}{ab} - \dfrac{2ab}{8} = $ _____

10. $\dfrac{xy}{5} + \dfrac{z}{3} = $ _____

11. $\dfrac{3y}{ab} - \dfrac{5y}{a} = $ _____

12. $\dfrac{x^2}{3z} + \dfrac{-x}{5z} = $ _____

13. $\dfrac{a}{b} - \dfrac{b}{a} = $ _____

14. $\dfrac{5}{p} - 1 = $ _____

15. $\dfrac{x+3}{y} + \dfrac{5}{x} =$ _____

16. $\dfrac{7}{x-3} - \dfrac{4}{x-3} =$ _____

17. $\dfrac{x-2}{x+2} - \dfrac{x+2}{x-2} =$ _____

18. $\dfrac{a+b}{a-b} + \dfrac{5}{a+b} =$ _____

19. $\dfrac{3}{x-2} + \dfrac{4}{x+3} =$ _____

20. $\dfrac{x-y}{x-2} - \dfrac{x+y}{x-2} =$ _____

21. $\dfrac{a^2}{4-a} + \dfrac{16}{a-4} =$ _____

22. $\dfrac{5}{(a-b)^2} - \dfrac{3}{(a-b)(a+b)} =$ _____

23. $\dfrac{16a}{xy} + \dfrac{8b}{yz} =$ _____

24. $\dfrac{p-q}{pq} - \dfrac{q-t}{qt} =$ _____

25. $\dfrac{1}{x-3} - \dfrac{1}{x+2} =$ _____

26. $\dfrac{a}{xy^2} + \dfrac{2}{xz^2} =$ _____

B2.4 MULTIPLICATIONS AND DIVISION OF RATIONAL EXPRESSIONS

Recall from arithmetic that to multiply fractions, simply find the product of the numerators and the denominators. After completing the multiplication, simplify the answer if possible. Exactly the same process is used in multiplying rational expressions.

Example A $\dfrac{3}{4} \cdot \dfrac{5}{6} = \dfrac{15}{24} = \dfrac{3 \cdot 5}{3 \cdot 8} = \dfrac{5}{8}.$

Example B $\dfrac{a}{x} \cdot \dfrac{b}{y} = \dfrac{ab}{xy}.$

The answer cannot be simplified.

Example C $\dfrac{ab^2}{10x} \cdot \dfrac{5b}{2a} = \dfrac{5ab^3}{20ax} = \dfrac{b^3}{4x}.$

This product was simplified.

There are times when factoring first may facilitate the multiplication process. Note in the following example that if you simply multiply the numerators and denominators first, you would have considerable simplifying to do before getting the answer. But by factoring, some of the terms cancel.

Example D $$\dfrac{x^2 - y^2}{x^2 + xy} \cdot \dfrac{x + y}{x^2 - 2xy + y^2} = \dfrac{(x - y)(x + y)(x + y)}{x(x + y)(x - y)(x - y)}$$

$$= \dfrac{x + y}{x(x - y)}$$

Dividing rational expressions also follows the same procedure as dividing numbers. The dividend is multiplied by the reciprocal of the divisor. $\dfrac{a}{b} \div \dfrac{c}{d} = \dfrac{a}{b} \cdot \dfrac{d}{c}.$ When a number or a rational expression is multiplied by its reciprocal, the product is 1.

Example E $\dfrac{2}{5} \cdot \dfrac{5}{2} = \dfrac{10}{10} = 1.$

$\dfrac{2}{5}$ and $\dfrac{5}{2}$ are reciprocals because their product equals 1.

Example F $\dfrac{ab}{3x} \cdot \dfrac{3x}{ab} = \dfrac{3abx}{3abx} = 1.$

Here $\dfrac{ab}{3x}$ and $\dfrac{3x}{ab}$ are reciprocals because their product is 1. **165**

This concept of reciprocals is now used to divide rational expressions. The general definition for division is given by

$$\frac{x}{y} \div \frac{a}{b} = \frac{x}{y} \cdot \frac{b}{a} = \frac{bx}{ay}$$

The dividend is multiplied by the reciprocal of the divisor. In this example, x, y, a, and b may be any number or rational expression, except y, a, and b cannot be 0.

Example G
$$\frac{8abx^2}{9z} \div \frac{4a^3xz}{3b}.$$

The reciprocal of the divisor is $\dfrac{3b}{4a^3xz}$.

$$\frac{8abx^2}{9z} \times \frac{3b}{4a^3xz} = \frac{24ab^2x^2}{36a^3xz^2}$$

$$= \frac{2b^2x}{3a^2z^2}$$

Example H
$$\frac{x}{x-1} \div \frac{x}{y}.$$

Reciprocal of $\dfrac{x}{y}$ is $\dfrac{y}{x}$.

$$\frac{x}{x-1} \cdot \frac{y}{x} = \frac{xy}{x(x-1)}$$

$$= \frac{y}{x-1}$$

When multiplying and dividing rational expressions, try to simplify as much as possible since this will facilitate computation and help minimize mistakes.

Example I
$$\frac{a^2-9}{111b^2} \div \frac{3-a}{37b}.$$

$$\frac{a^2-9}{111b^2} \div \frac{3-a}{37b} = \frac{a^2-9}{111b^2} \times \frac{37b}{3-a}$$

$$= \frac{(a-3)(a+3)37b}{37b(3b)(3-a)}$$

$$= \frac{-(a+3)}{3b}$$

$$= -\frac{a+3}{3b}$$

B2.4 PROBLEMS

Solve.

1. $\dfrac{2x}{10} \cdot \dfrac{5}{4x} = $ _____

2. $\dfrac{3a}{x} \cdot \dfrac{b}{6a} = $ _____

3. $\dfrac{ab}{xy} \cdot \dfrac{xy}{ab} = $ _____

4. $\dfrac{5x}{-2} \cdot \dfrac{-4y}{xz} = $ _____

5. $\dfrac{4a^2}{8x} \cdot \dfrac{2x^3}{6a} = $ _____

6. $\dfrac{-9r}{4t} \cdot \dfrac{2t^2}{3r} = $ _____

7. $\dfrac{a}{-b} \cdot \dfrac{-a^2b^2}{cd} = $ _____

8. $\dfrac{-28x^2}{-3y} \cdot \dfrac{7y^2}{-4x} = $ _____

9. $\dfrac{2x+6}{2x} \cdot \dfrac{x}{6} = $ _____

10. $\dfrac{5}{q+2} \cdot \dfrac{(q+2)^2}{10} = $ _____

11. $\dfrac{x+5}{4} \cdot \dfrac{2-y}{x+5} = $ _____

12. $\dfrac{7}{4+x} \cdot \dfrac{x+4}{7} = $ _____

13. $\dfrac{3a+9}{a+3} \cdot \dfrac{2a+3}{a+3} = $ _____

14. $\dfrac{a^2-a-12}{2-a} \cdot \dfrac{a^2+4a-12}{a+3} = $ _____

15. $\dfrac{4y^2}{4(a+b)} \cdot \dfrac{4a+4b}{y(a+b)} = $ _____

16. $\dfrac{a+b}{a-2} \cdot \dfrac{a^2-4}{a+2} = $ _____

17. $\dfrac{(a+b)^2}{a-b} \cdot \dfrac{(a-b)^2}{2(a+b)} = $ _____

18. $\dfrac{w^2}{4w-4y} \cdot \dfrac{w^2-y^2}{w(w+y)} = $ _____

19. $\dfrac{a^2-9}{a+3} \cdot \dfrac{6}{a-3} = $ _____

167

20. $\dfrac{a}{x} \div \dfrac{b}{y} = $ _____

21. $\dfrac{2x}{3y} \div \dfrac{4x}{9y} = $ _____

22. $\dfrac{ab}{4} \div \dfrac{a^2 b}{8} = $ _____

23. $\dfrac{x}{2y} \div \dfrac{x}{-8y} = $ _____

24. $\dfrac{7ab^2}{ab} \div \dfrac{a^2 b}{7ab} = $ _____

25. $\dfrac{5pq^2}{xy} \div \dfrac{4}{10xy} = $ _____

26. $\dfrac{c^2 d}{x} \div \dfrac{d}{cx} = $ _____

27. $\dfrac{a^2 - b^2}{b - a} \div \dfrac{a + b}{b - a} = $ _____

28. $\dfrac{4ab}{10xy} \div \dfrac{8a^2 b^2}{5x} = $ _____

29. $\dfrac{5(a^2 - 4b^2)}{a + b} \div \dfrac{a - 2b}{10(a + b)} = $ _____

30. $\dfrac{25}{xy} \div \dfrac{75x}{30y} = $ _____

31. $\dfrac{3xyz}{x^2 - y^2} \div \dfrac{4z}{3x - 3y} = $ _____

32. $\dfrac{5z - 5}{4w} \div \dfrac{z - 1}{5w} = $ _____

33. $\dfrac{a(b + c)^2}{a - c} \div \dfrac{a + b}{a - c} = $ _____

34. $\dfrac{a^2 - b^2}{25a} \div \dfrac{a^2 - 2ab + b^2}{50a} = $ _____

35. $\dfrac{b(a^2 - b^2)}{11a} \div \dfrac{b}{44a} = $ _____

36. $\dfrac{x^2 + x - 6}{x - 3} \div \dfrac{x + 3}{x - 3} = $ _____

B2.5 SIMPLIFICATION OF A COMPOUND RATIONAL EXPRESSION

Much of algebra is devoted to the various processes of simplifying expressions. This is essential because it greatly reduces the amount of computation that may be necessary, and it also enhances understanding the relationships that exist.

We have dealt with several kinds of rational expressions, both in simplifying them and performing operations on them. We now tie many of these techniques together. A compound rational expression is one that has fractions and/or rational expressions in the denominator or numerator, or both. Simplifying them means to operate on them so they result in a single rational expression with no fractions in the numerator or denominator.

Example A Simplify $\dfrac{a}{a + \dfrac{1}{2}}$.

The challenge is to remove the fraction from the denominator in this rational expression. Begin with the denominator because it contains a rational expression.

$$a + \frac{1}{2} = \frac{2a + 1}{2}$$

Now replace this in the original rational expression and divide.

$$\frac{a}{\dfrac{2a + 1}{2}} = \frac{a}{1} \div \frac{2a + 1}{2} = \frac{a}{1} \times \frac{2}{2a + 1} = \frac{2a}{2a + 1}$$

Example B Simplify $\dfrac{\dfrac{x}{y} - 8}{z}$.

This time, begin in the numerator because it contains a rational expression.

$$\frac{x}{y} - 8 = \frac{x - 8y}{y}$$

Now put this back into the original expression and divide.

$$\frac{\dfrac{x - 8y}{y}}{z} = \frac{x - 8y}{y} \cdot \frac{1}{z}$$

$$= \frac{x - 8y}{yz}$$

In the case where both the numerator and the denominator contain rational expressions, do each separately and then divide.

Example C   Simplify $\dfrac{1 + \dfrac{3}{a}}{2 + \dfrac{3}{b}}$.

First, the numerator:

$$1 + \frac{3}{a} = \frac{a + 3}{a}$$

Next, the denominator:

$$2 + \frac{3}{b} = \frac{2b + 3}{b}$$

Then put them together:

$$\frac{\dfrac{a + 3}{a}}{\dfrac{2b + 3}{b}} = \frac{a + 3}{a} \cdot \frac{b}{2b + 3} = \frac{b(a + 3)}{a(2b + 3)}$$

Example D $\dfrac{\dfrac{4}{x} + 1}{\dfrac{4}{x} + x} = \dfrac{\dfrac{4 + x}{x}}{\dfrac{4 + x^2}{x}} = \dfrac{4 + x}{x} \cdot \dfrac{x}{4 + x^2} = \dfrac{x(x + 4)}{x(x^2 + 4)} = \dfrac{x + 4}{x^2 + 4}$

Example E $\dfrac{\dfrac{x^2}{4} - 9}{\dfrac{3}{2} + \dfrac{x}{4}} = \dfrac{\dfrac{x^2 - 36}{4}}{\dfrac{12 + 2x}{8}}$

$$= \frac{x^2 - 36}{4} \times \frac{8}{12 + 2x}$$

$$= \frac{8(x - 6)(x + 6)}{8(x + 6)}$$

$$= x - 6$$

B2.5 PROBLEMS

Simplify.

1. $\dfrac{a}{\dfrac{a}{2}+1} = $ _____

2. $\dfrac{2x-3}{\dfrac{x}{y}+3} = $ _____

3. $\dfrac{\dfrac{5b-c}{a}}{\dfrac{b}{a}+5} = $ _____

4. $\dfrac{\dfrac{x}{x+y}-1}{\dfrac{x}{x+y}} = $ _____

5. $\dfrac{b}{b-\dfrac{3}{4}} = $ _____

6. $\dfrac{\dfrac{a}{b}+4}{\dfrac{a}{b}-4} = $ _____

7. $\dfrac{\dfrac{2-x}{y}+1}{y^2} = $ _____

8. $\dfrac{\dfrac{-3+x}{x-3}}{x}-2 = $ _____

9. $\dfrac{\dfrac{a^2-b^2}{x}}{\dfrac{a}{x}-b} = $ _____

171

10. $\dfrac{\dfrac{7-y}{y-7}}{y+\dfrac{7}{y}} =$ _____

11. $\dfrac{2+\dfrac{5}{a}}{a+\dfrac{5}{a}} =$ _____

12. $\dfrac{\dfrac{a-b}{4}+4}{16} =$ _____

13. $\dfrac{\dfrac{x^2-y^2}{x+y}-6}{\dfrac{x-y}{x+y}} =$ _____

14. $\dfrac{\dfrac{a-b}{a+b}-\dfrac{2}{3}}{\dfrac{2}{3}} =$ _____

15. $\dfrac{\dfrac{y}{x}-\dfrac{x}{y}}{\dfrac{y}{x}+\dfrac{x}{y}} =$ _____

16. $\dfrac{\dfrac{a^2-2ab+b^2}{a-b}}{\dfrac{a-b}{ab}} =$ _____

17. $\dfrac{\dfrac{4x^2-9y^2}{xy}}{\dfrac{2x-3y}{xy}} =$ _____

18. $\dfrac{\dfrac{a-b}{a^2-b^2}-5}{\dfrac{a-b}{a+b}} =$ _____

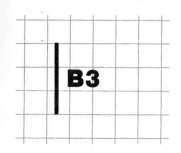

B3 | OPERATIONS WITH INTEGRAL EXPONENTS

B3.1 DEFINITION OF EXPONENTIATION WITH POSITIVE EXPONENTS

You have already learned some ideas about exponents, and you have worked with them. In this chapter, you will continue your exposure, even with 0 and negative exponents. But in this first unit, we'll restrict ourselves to positive whole number exponents.

The definition is:

x^1 means x

x^2 means $x \cdot x$

x^3 means $x \cdot x \cdot x$

x^4 means $x \cdot x \cdot x \cdot x$

x^5 means $x \cdot x \cdot x \cdot x \cdot x$

and so on.

The expression x^n for n a positive integer is defined as follows:

$$x^n = \underbrace{x \cdot x \cdot x \cdots x}_{n \text{ factors}}$$

The expression x^n is called a *power*. The number x is the *base*. The *exponent n* is the number of times the base is used as a factor.

Example A   What does 2^3 mean?

2^3 means $2 \cdot 2 \cdot 2$, or 8. Thus, $2^3 = 8$.

181

Example B   What does 3^2 mean?

3^2 means $3 \cdot 3$, or 9. Thus, $3^2 = 9$. Notice that $2^3 \neq 3^2$.

Example C   Is 27 a power?

$27 = 3 \cdot 9 = 3 \cdot 3 \cdot 3 = 3^3$, so it is a power.

Example D   Is 64 a power?

Yes, because $64 = 8 \cdot 8 = 8^2$. But also $64 = 4 \cdot 4 \cdot 4 = 4^3$. Furthermore, $64 = 2 \cdot 2 \cdot 2 \cdot 2 \cdot 2 \cdot 2 = 2^6$. Thus, 64 is a power in several factors. You will learn in the next unit how all these relate to one another.

Example E   Is 12 a power?

No. The number 12 factors as $3 \cdot 4 = 3 \cdot 2 \cdot 2$. One base is not repeated, so it is not a power.

Example F   If a power is multiplied by a power, do you get a power? Sometimes yes, sometimes no. Here are some cases to study.

a. $3^2 \cdot 3^3 = (3 \cdot 3)(3 \cdot 3 \cdot 3) = 3 \cdot 3 \cdot 3 \cdot 3 \cdot 3 = 3^5$
In this case, the answer is yes.

b. $2^2 \cdot 3^3 = (2 \cdot 2)(3 \cdot 3 \cdot 3) = 2 \cdot 2 \cdot 3 \cdot 3 \cdot 3$
In this case, the answer is no.

c. $2^2 \cdot 4^2 = (2 \cdot 2)(4 \cdot 4) = 2 \cdot 2 \cdot 4 \cdot 4$
$= 2 \cdot 2 \cdot 2 \cdot 2 \cdot 2 \cdot 2 = 2^6$
In this case, the answer is yes.

d. $2^3 \cdot 3^3 = (2 \cdot 2 \cdot 2)(3 \cdot 3 \cdot 3)$
$= 2 \cdot 2 \cdot 2 \cdot 3 \cdot 3 \cdot 3$
$= (2 \cdot 3)(2 \cdot 3)(2 \cdot 3)$
$= 6 \cdot 6 \cdot 6 = 6^3$
In this case, the answer is yes again.

You see by Example F that one must be careful in dealing with powers and exponents. But we will now present the first law of exponents. Examples will lead the way.

Example G   Can $5^4 \cdot 5^3$ be written as a power?

Yes, because $5^4 \cdot 5^3 = (5 \cdot 5 \cdot 5 \cdot 5)(5 \cdot 5 \cdot 5)$
$= 5 \cdot 5 \cdot 5 \cdot 5 \cdot 5 \cdot 5 \cdot 5$
$= 5^7$.

Example H Can $x^m \cdot x^n$ be written as a power?

Yes.

$$x^m \cdot x^n = \underbrace{(x \cdot x \cdot x \cdots x)}_{m \text{ factors}} \underbrace{(x \cdot x \cdot x \cdots x)}_{n \text{ factors}}$$

$$= \underbrace{x \cdot x \cdot x \cdots x}_{m + n \text{ factors}}$$

$$= x^{m+n}.$$

This is, then, the

FIRST LAW OF EXPONENTS

$x^m \cdot x^n = x^{m+n}$

To multiply powers with the same base, add the exponents.

Example I Is it true that $x^m \cdot y^n = (xy)^{m+n}$?

No! Here is an example.

$$2^3 \cdot 3^2 = (2 \cdot 2 \cdot 2)(3 \cdot 3)$$

$$= 8 \cdot 9$$

$$= 72$$

$$(2 \cdot 3)^{3+2} = 6^5$$

$$= 6 \cdot 6 \cdot 6 \cdot 6 \cdot 6$$

$$= 7776$$

Remember: The first law of exponents *demands* that the base be the same. Then to multiply, you add the exponents.

Example J Use the first law of exponents to find $5^2 \cdot 5$.

$$5^2 \cdot 5 = 5^2 \cdot 5^1$$

$$= 5^{2+1}$$

$$= 5^3$$

$$= 125$$

Example K Use the first law of exponents to find $(ab)(bc)(cd)$.

$$(ab)(bc)(cd) = a^1 \cdot b^1 \cdot b^1 \cdot c^1 \cdot c^1 \cdot d^1$$

$$= a^1 \cdot b^{1+1} \cdot c^{1+1} \cdot d^1$$

$$= ab^2c^2d$$

Example L　Use the first law of exponents to find $(x^2y)(y^3z)(z^3)$.

$$(x^2y)(y^3z)(z^3) = x^2y^{1+3}z^{1+3}$$
$$= x^2y^4z^4$$

Example M　Use the first law of exponents to find $(-3)^3(-3)^2$.

$$(-3)^3(-3)^2 = (-3)^5 = -243$$

Example N　Simplify $-5x(4x^2)$

$$-5x(4x^2) = -20x^3$$

Example O　Use exponents to write $xx - yyy$.

$$xx - yyy = x^2 - y^3$$

Example P　Use exponents to write $5^2(5xy)(x^2y) - 3(3^4x)(x^5y^2)$.

$$5^2(5xy)(x^2y) - 3(3^4x)(x^5y^2) = 5^3x^3y^2 - 3^5x^6y^2$$

B3.1 PROBLEMS

Write in standard form (i.e., without exponents).

1. $2^5 =$ _____

2. $3^4 =$ _____

3. $(-4)^3 =$ _____

4. $(-5)^3 =$ _____

5. $(-3)^2 \cdot 4^1 =$ _____

6. $(-2)^3(-3)^1 =$ _____

7. $-5^3 =$ _____

8. $-4^3 =$ _____

9. $-3(-2)^3 =$ _____

10. $-2(-3)^3 =$ _____

Use exponents to write each expression.

11. $aaaaa =$ _____

12. $bbbbbb =$ _____

13. $aaabbbb =$ _____

14. $bbccccc =$ _____

15. $3xxxyy =$ _____

16. $5uuvvvv =$ _____

17. $7pppp - qqq =$ _____

18. $mmm + 3nnn =$ _____

19. $-3xxy + 13xyy = $ _____

20. $-7aaab - 5abb = $ _____

Use the first law of exponents to simplify. Use exponents in your answers.

21. $7^5 \cdot 7^8 = $ _____

22. $5^7 \cdot 5^9 = $ _____

23. $x^4 \cdot x^6 = $ _____

24. $y^3 \cdot y^8 = $ _____

25. $-a \cdot a^5 = $ _____

26. $b^2(-b)^3 = $ _____

27. $m^2(-m)^4 = $ _____

28. $-n \cdot n^7 = $ _____

29. $u^2 \cdot u^3 \cdot u^4 = $ _____

30. $v^3 \cdot v^2 \cdot v = $ _____

31. $(a^2b)(-a^2)(-b)^2 = $ _____

32. $(xy^3)(-x)^3(-y^2) = $ _____

33. $a^2(a^4b^3) = $ _____

34. $(x^3y^3)x^4 = $ _____

35. $(-s)^3(-s)^2(-s) = $ _____

36. $(-x)^2(-x)^4(-x)^6 = $ _____

37. $(-s^3)(-s^2)(-s) = $ _____

38. $(-x^2)(-x^4)(-x^6) = $ _____

B3.2 LAWS OF EXPONENTS WITH POSITIVE EXPONENTS

You have already learned about and used the first law of exponents. We repeat it here for the sake of completeness.

> ### FIRST LAW OF EXPONENTS
>
> $$x^m \cdot x^n = x^{m+n}$$
>
> To multiply powers with the same base, add the exponents.

This unit will cover two more laws of exponents. The second law of exponents is suggested by the following examples.

Example A Does $(ab)^3 = a^3 b^3$?

Let's see if it does. By the definition of exponentiation,

$$(ab)^3 = (ab)(ab)(ab)$$
$$= a \cdot a \cdot a \cdot b \cdot b \cdot b$$
$$= a^3 b^3$$

The answer is yes.

Example B Does $(xy)^5 = x^5 y^5$?

Yes, because

$$(xy)^5 = (xy)(xy)(xy)(xy)(xy)$$
$$= x \cdot x \cdot x \cdot x \cdot x \cdot y \cdot y \cdot y \cdot y \cdot y$$
$$= x^5 y^5$$

Example C Use Example B to calculate $2^5 \cdot 5^5$.

$2^5 = 32$ and $5^5 = 3125$. To multiply these is a big task. But by Example B,

$$2^5 \cdot 5^5 = (2 \cdot 5)^5$$
$$= 10^5$$
$$= 100{,}000$$

These examples are special cases of the second law of exponents.

Example D   Write $(xy)^n$ as a product of powers.

We proceed as before.

$$(xy)^n = \underbrace{(xy)(xy)(xy) \cdots (xy)}_{n \text{ factors of } xy}$$

$$= \underbrace{x \cdot x \cdot x \cdots x}_{n \text{ factors of } x} \cdot \underbrace{y \cdot y \cdot y \cdots y}_{n \text{ factors of } y}$$

$$= x^n y^n$$

> SECOND LAW OF EXPONENTS
>
> $$(xy)^n = x^n y^n$$
>
> An exponent applied to a product yields the same result when it is applied to each factor.

Example E   Write $(xyz)^n$ as a product of powers.

By the second law of exponents, $(xyz)^n = (xy)^n z^n$. Now apply the second law to $(xy)^n$ to yield

$$(xyz)^n = (xy)^n z^n = x^n y^n z^n$$

The first two laws of exponents can be combined to suggest the third law of exponents.

Example F   Use the second law of exponents to write $(x^3)^5$ as a power.

By the definition of x^3,

$$(x^3)^5 = (x \cdot x \cdot x)^5$$

Now by the second law of exponents applied as in Example E,

$$(x \cdot x \cdot x)^5 = x^5 x^5 x^5$$

Now by the first law of exponents,

$$x^5 x^5 x^5 = x^{5+5+5} = x^{15}$$

The net result is

$$(x^3)^5 = x^{15} = x^{3 \cdot 5}$$

Example G   Use the second law of exponents to write $(a^4)^3$ as a power.

We will go through the steps without comment. Be sure you understand each step.

$$
\begin{aligned}
(a^4)^3 &= (a \cdot a \cdot a \cdot a)^3 \\
&= a^3 \cdot a^3 \cdot a^3 \cdot a^3 \\
&= a^{3+3+3+3} \\
&= a^{12} \\
&= a^{4 \cdot 3}
\end{aligned}
$$

Example H   Write $(x^m)^n$ as a power.

$$
\begin{aligned}
(x^m)^n &= \underbrace{(x \cdot x \cdot x \cdots x)^n}_{m \text{ factors of } x^n} \\
&= x^{n+n+n+ \cdots +n} \\
&= x^{mn}
\end{aligned}
$$

THIRD LAW OF EXPONENTS

$$(x^m)^n = x^{mn}$$

To apply an exponent to a power, multiply the exponents.

Example I   Simplify and identify the laws used.

$$
\begin{aligned}
x^4(x^3y)^2 &= x^4(x^3)^2y^2 & \text{Second law} \\
&= x^4x^6y^2 & \text{Third law} \\
&= x^{4+6}y^2 & \text{First law} \\
&= x^{10}y^2
\end{aligned}
$$

There's only one more item to take care of in this unit: division with exponents.

Example J   Simplify $\dfrac{x^5}{x^2}$.

By the first law of exponents, $x^2 \cdot x^3 = x^5$. Thus,

$$\frac{x^5}{x^2} = x^3$$

Another way to do this is to think of cancellation.

$$\frac{x^5}{x^2} = \frac{\cancel{x} \cdot \cancel{x} \cdot x \cdot x \cdot x}{\cancel{x} \cdot \cancel{x}}$$

$$= x^3$$

We'll do more with division of exponents in the next sections.

Example K   Simplify $[(3x^2y)^2]^3$

$$\left[(3x^2y)^2\right]^3 = (9x^4y^2)^3$$

$$= 729x^{12}y^6$$

B3.2 PROBLEMS

Simplify each expression. If the product of any numbers involved is 100 or less, then write it in standard form; otherwise, write the numbers with appropriate exponents. You do not have to identify which laws you use.

1. $[(-2)(-3)]^2 = $ _____

2. $-[2(-5)]^2 = $ _____

3. $(-2)2^3 = $ _____

4. $(-3)(-3)^2 = $ _____

5. $(6 \cdot 7)^5 = $ _____

6. $(5^4)^5 = $ _____

7. $(13^3)^4 = $ _____

8. $(8 \cdot 9)^3 \cdot 8 = $ _____

9. $(-2x)^2 = $ _____

10. $-(3y)^4 = $ _____

11. $-(2x)^2 = $ _____

12. $(-3y)^4 = $ _____

13. $3x^4 \cdot 2x^5 = $ _____

14. $(-2m)^3(-3m)^2 = $ _____

15. $(-4k)^2(-2k)^3 = $ _____

16. $8t^3 \cdot 9t^6 = $ _____

17. $(a^2 \cdot a)^3 = $ _____

18. $(-x^2)(-x)^2x = $ _____

19. $(-t)^3(-t^3)t = $ _____

20. $(b \cdot b^3)^2 = $ _____ **191**

21. $(r^2s^3r)^4 = $ _____

22. $(-x)(-y^2)^2(-z^3)^3 = $ _____

23. $(-a^2)^3(-b)^2(-c)^4 = $ _____

24. $(ab^4a^5)^2 = $ _____

25. $[(pq)^2]^4 = $ _____

26. $(-x^2)^3 = $ _____

27. $(-u^3)^4 = $ _____

28. $[(mn^2)^3]^3 = $ _____

29. $[(3xy^2z^3)^2]^3 = $ _____

30. $[(-x)^3]^2 = $ _____

31. $(-u^4)^3 = $ _____

32. $[(2a^3b^2c)^3]^2 = $ _____

33. $\dfrac{3^6}{3^2} = $ _____

34. $\dfrac{5^7}{5^5} = $ _____

35. $\dfrac{x^4}{x} = $ _____

36. $\dfrac{u^6v^4}{u^2v^2} = $ _____

37. $\dfrac{(-m)^3n^4}{-mn^2} = $ _____

38. $\dfrac{a^5}{a} = $ _____

39. $\dfrac{(a^3b^2)^2}{(ab)^2} = $ _____

40. $\dfrac{p^5(-q)^4}{-p^2q} = $ _____

B3.3 SIMPLIFICATION OF AN EXPRESSION WITH POSITIVE EXPONENTS

We now continue the activities of the last unit in terms of simplifying expressions involving exponents.

Example A   Simplify $(a^3b^2)^2(ab)^3$.

First, by the second law of exponents,

$$(a^3b^2)^2(ab)^3 = (a^3)^2(b^2)^2a^3b^3$$

Now use the third law:

$$= a^6b^4a^3b^3$$

Now use the first law:

$$= a^{6+3}b^{4+3} = a^9b^7$$

Example B   Simplify $(-3x^4yz^2)^3(2x^2z)^4$.

This time, the simplification will be done without explicit reference to which law is being used. However, follow each step closely.

$$(-3x^4yz^2)^3(2x^2z)^4 = (-3)^3(x^4)^3y^3(z^2)^3(2)^4(x^2)^4z^4$$

$$= -27x^{12}y^3z^6 \cdot 16x^8z^4$$

$$= -432x^{12+8}y^3z^{6+4} = -432x^{20}y^3z^{10}$$

Example C   Simplify $(-2pq^2r^3)^5(-3p^3qr)^6$.

$$(-2pq^2r^3)^5(-3p^3qr)^6 = -2^53^6p^{23}q^{16}r^{21}$$

Example D   Simplify $\dfrac{6x^7}{2x^3}$.

One way to proceed is to rewrite this in terms that are easy to cancel.

$$\frac{6x^7}{2x^3} = \frac{\cancel{2} \cdot 3 \cdot x \cdot x \cdot x \cdot x \cdot \cancel{x} \cdot \cancel{x} \cdot \cancel{x}}{\cancel{2} \cdot \cancel{x} \cdot \cancel{x} \cdot \cancel{x}} = 3 \cdot x \cdot x \cdot x \cdot x = 3x^4$$

Example E   Simplify $\dfrac{3a^2}{9a^5}$.

Again,

$$\frac{3a^2}{9a^5} = \frac{\cancel{3} \cdot \cancel{a} \cdot \cancel{a}}{\cancel{3} \cdot 3 \cdot a \cdot a \cdot a \cdot \cancel{a} \cdot \cancel{a}} = \frac{1}{3 \cdot a \cdot a \cdot a} = \frac{1}{3a^3}$$

Example F   Simplify $\dfrac{4m^4n}{6m^2n^3}$.

$$\frac{4m^4n}{6m^2n^3} = \frac{\cancel{2} \cdot 2 \cdot \cancel{m} \cdot \cancel{m} \cdot m \cdot m \cdot \cancel{n}}{\cancel{2} \cdot 3 \cdot \cancel{m} \cdot \cancel{m} \cdot n \cdot n \cdot \cancel{n}} = \frac{2 \cdot m \cdot m}{3 \cdot n \cdot n} = \frac{2m^2}{3n^2}$$

Example G   Simplify $\dfrac{2r^3s^2}{8rs^3}$.

See if you can understand the answer without intermediate steps.

$$\frac{2r^3s^2}{8rs^3} = \frac{r^2}{4s}$$

Example H   Simplify $\dfrac{(-2)^3u^5v^2}{(-6)^2u^2v^7}$.

When negative signs are involved, be extra careful.

$$\frac{(-2)^3u^5v^2}{(-6)^2u^2v^7} = \frac{-8u^5v^2}{36u^2v^7} = -\frac{2u^3}{9v^5}$$

Example I   What is meant by $\left(\dfrac{2}{3}\right)^4$?

It means $\dfrac{2}{3}$ is used as a factor 4 times. Thus,

$$\left(\frac{2}{3}\right)^4 = \frac{2}{3} \cdot \frac{2}{3} \cdot \frac{2}{3} \cdot \frac{2}{3} = \frac{2 \cdot 2 \cdot 2 \cdot 2}{3 \cdot 3 \cdot 3 \cdot 3} = \frac{16}{81}$$

Example J   Rewrite $\left(\dfrac{x}{y}\right)^n$.

$$\left(\frac{x}{y}\right)^n = \underbrace{\frac{x}{y} \cdot \frac{x}{y} \cdot \frac{x}{y} \cdots \frac{x}{y}}_{n \text{ factors}}$$

$$= \frac{x \cdot x \cdot x \cdots x}{y \cdot y \cdot y \cdots y} = \frac{x^n}{y^n}$$

Example K   Simplify $\left(\dfrac{3a^4}{b}\right)\left(\dfrac{2a}{b}\right)^3$.

$$\left(\frac{3a^4}{b}\right)\left(\frac{2a}{b}\right)^3 = \frac{3a^4}{b} \cdot \frac{(2a)^3}{b^3}$$

$$= \frac{3a^4}{b} \cdot \frac{2^3a^3}{b^3} = \frac{3a^4}{b} \cdot \frac{8a^3}{b^3} = \frac{24a^7}{b^4}$$

B3.3 PROBLEMS

Simplify each expression.

1. $(ab)^3 a^2 =$ _____

2. $(u^2 v)^2 (uv^3) =$ _____

3. $(mn^2)^3 (m^2 n) =$ _____

4. $y^3 (xy)^2 =$ _____

5. $(r^2 s)^2 (st)^3 =$ _____

6. $-(pq)^2 (-p)^3 =$ _____

7. $(-uv)^2 (-u^2)^3 =$ _____

8. $(ab^3)^2 (ab)^2 =$ _____

9. $(-2bc^2)^3 (3b^3 c)^2 =$ _____

10. $(-3v^2 w)^2 (2vw^2)^3 =$ _____

11. $(5x^3 y^2)^4 (7xy^5)^3 =$ _____

12. $(-c^5 d^2)^3 (-c^2 d^3)^5 =$ _____

13. $(-st^2)(-t^3 u^4)^2 (-u^5 v^6)^3 =$ _____

14. $(xy^2 z)(x^2 yz)^2 (xyz^2)^3 =$ _____

15. $\dfrac{4p^5}{2p^2} = $ _____

16. $\dfrac{3r^4}{2r} = $ _____

17. $\dfrac{2q^3}{3q} = $ _____

18. $\dfrac{6b^6}{3b^4} = $ _____

19. $\dfrac{r^6 s^4}{r^3 s^2} = $ _____

20. $\dfrac{-u^6 v^5}{\left(-u^2 v\right)^2} = $ _____

21. $\dfrac{\left(-c^3 d^2\right)^3}{-c^5 d^3} = $ _____

22. $\dfrac{p^7 q^5}{p^4 q^3} = $ _____

23. $\dfrac{12a^5}{6a^7} = $ _____

24. $\dfrac{3x^6}{9x^{10}} = $ _____

25. $\dfrac{5b^3}{15b^8} = $ _____

26. $\dfrac{8w^4}{2w^9} = $ _____

27. $\dfrac{10x^5 y^2}{12x^2 y^4} = $ _____

28. $\dfrac{\left(-2r^2 s^3\right)^3}{-6r^7 s^5} = $ _____

29. $\dfrac{-36m^5 n^7}{\left(-3m^4 n^2\right)^2} = $ _____

30. $\dfrac{15a^7 b^3}{20a^4 b^8} = $ _____

31. $\left(\dfrac{3u^2}{5v^3}\right)^2 = $ _____

32. $\left(\dfrac{2x^3}{3y^2}\right)^3 = $ _____

33. $\left(\dfrac{x^2}{y}\right)\left(\dfrac{x}{y^2}\right)^3 = $ _____

34. $\left(\dfrac{u}{v^3}\right)^2\left(\dfrac{u^3}{v}\right) = $ _____

35. $\left(\dfrac{p}{q^3}\right)^2\left(\dfrac{q^2}{p^3}\right)^3 = $ _____

36. $\left(\dfrac{c^3}{d^2}\right)^4\left(\dfrac{d^3}{c^2}\right)^2 = $ _____

37. $\left(\dfrac{a}{b^2}\right)\left(\dfrac{2b^3 c^2}{a}\right)^2\left(\dfrac{ab}{3c}\right)^3 = $ _____

38. $\left(\dfrac{w^3}{x^2}\right)\left(\dfrac{3x^2 y^2}{w^2}\right)\left(\dfrac{x}{2y^2}\right)^3 = $ _____

B3.4 DEFINITION OF EXPONENTIATION WITH INTEGRAL EXPONENTS

Thus far, you have been dealing with powers where the exponent is a positive integer, such as

$a^5, 2^7, 3^1x^8, 5^4t^3u^6v$, and so on.

Also, you are aware of some laws of exponents.

The purpose of this unit is to expand the idea of exponentiation so that 0 and negative integers can serve as exponents. As a result, you will know by the end of this unit what is meant by expressions such as

$a^{-5}, 2^{-7}, 3^{-1}x^8, 5^4t^3u^{-6}v^0$, and so on.

In the next two units, you will learn more about laws of exponents and simplifying expressions involving exponents.

Definition of Zero Exponent

In trying to develop a meaning for an exponent of 0, we shall keep the rules for operating with exponents intact. Some familiar examples will pave the way.

Example A   Simplify $\dfrac{x^7}{x^3}$.

$$\frac{x^7}{x^3} = \frac{x \cdot x \cdot x \cdot x \cdot \cancel{x} \cdot \cancel{x} \cdot \cancel{x}}{\cancel{x} \cdot \cancel{x} \cdot \cancel{x}}$$

$$= x \cdot x \cdot x \cdot x = x^4 = x^{7-3}$$

Example B   Simplify $\dfrac{a^5}{a^8}$.

$$\frac{a^5}{a^8} = \frac{\cancel{a} \cdot \cancel{a} \cdot \cancel{a} \cdot \cancel{a} \cdot \cancel{a}}{a \cdot a \cdot a \cdot \cancel{a} \cdot \cancel{a} \cdot \cancel{a} \cdot \cancel{a} \cdot \cancel{a}}$$

$$= \frac{1}{a \cdot a \cdot a} = \frac{1}{a^3} = \frac{1}{a^{8-5}}$$

In each example, notice that we got the result by subtraction of the smaller exponent from the larger one. The result goes in the numerator or the denominator as appropriate.

Example C   Using the previous examples, let us define x^0?

$\dfrac{x^5}{x^5} = x^{5-5} = x^0$. Also, $\dfrac{x^5}{x^5} = 1$. Notice that if $x = 0$, then $\dfrac{x^5}{x^5}$ is not a number.

In general, $\dfrac{x^n}{x^n} = x^0$ and $\dfrac{x^n}{x^n} = 1$.

| DEFINITION
| If $x \neq 0$, then $x^0 = 1$.

197

Example D Does this definition work with the first law of exponents?

Yes: $x^0 \cdot x^n = x^{0+n} = x^n$ and, by the definition, $x^0 \cdot x^n = 1 \cdot x^n = x^n$.

Definition of Negative Exponents

We approach the idea of negative exponents in the same way: by insisting that our laws of exponents be kept intact.

Example E If we keep the first law of exponents, how should 2^{-3} be defined?

Since $2^m \cdot 2^n = 2^{m+n}$, then $2^3 \cdot 2^{-3}$ should be $2^{3-3} = 2^0 = 1$. But $2^3 \cdot \dfrac{1}{2^3} = 1$. So we should want to define 2^{-3} to be $\dfrac{1}{2^3}$.

Example F Use this idea to simplify $\dfrac{a^5}{a^8}$.

$\dfrac{a^5}{a^8} = a^{5-8} = a^{-3} = \dfrac{1}{a^3}$. Compare this with Example B.

Example G Does 0^{-4} make any sense?

No. If this idea were to work, then 0^{-4} would be $\dfrac{1}{0^4}$, which is not a number.

> DEFINITION
> If $x \neq 0$ and if n is a positive integer, then $x^{-n} = \dfrac{1}{x^n}$.

Example H Does this definition work with the first law of exponents?

Yes: $x^n \cdot x^{-n} = x^{n-n} = x^0 = 1$ and, by the definition, $x^n \cdot \dfrac{1}{x^n} = 1$.

Example I Evaluate $\dfrac{3}{2^{-4}}$.

Since $2^{-4} = \dfrac{1}{2^4}$, then $\dfrac{1}{2^{-4}} = 2^4$. Thus, $\dfrac{3}{2^{-4}} = 3 \cdot 2^4 = 3 \cdot 16 = 48$.

Example J Write $\dfrac{5x^3 y^{-2}}{z^{-4}}$ without using negative exponents.

$$\frac{5x^3 y^{-2}}{z^{-4}} = \frac{5x^3 z^4}{y^2}$$

Example K Simplify $\dfrac{3x^2 y^{-2}}{x^{-3} y}$.

$$\frac{3x^2 y^{-2}}{x^{-3} y} = \frac{3x^2 x^3}{y^2 y} = \frac{3x^5}{y^3}$$

B3.4 PROBLEMS

Evaluate each expression in the usual way when it is defined. If it is not defined, write N.D.

1. $5^{-2} = $ _____

2. $3^{-1} = $ _____

3. $2^{-1} = $ _____

4. $4^{-3} = $ _____

5. $2^3 \cdot 3^{-2} = $ _____

6. $7^0 = $ _____

7. $\dfrac{(8-8)^0}{4^0} = $ _____

8. $4^{-3} \cdot 3^3 = $ _____

9. $8^0 = $ _____

10. $6^0(5-5)^0 = $ _____

11. $\left(\dfrac{1}{2} - \dfrac{1}{4} - \dfrac{1}{4} \right)^{-3} = $ _____

12. $(6-9)^{4-3-1} = $ _____

13. $(4-7)^{3-1-2} = $ _____

14. $\left(\dfrac{1}{3} - \dfrac{1}{6} - \dfrac{1}{6} \right)^{-4} = $ _____

15. $(-3)^{-4} = $ _____

16. $-(-2)^{-5} = $ _____

199

Write each expression without using negative exponents, and simplify where possible. Where numbers are involved, write the numbers in standard form.

17. $xy^{-2} =$ _____

18. $2^{-3}m =$ _____

19. $3^{-4}k =$ _____

20. $a^{-3}b =$ _____

21. $2^{-1} + 4^{-1} =$ _____

22. $3^{-1} - 6^{-1} =$ _____

23. $(2 + 4)^{-1} =$ _____

24. $(3 - 6)^{-1} =$ _____

25. $3^{-2}x^{-3} =$ _____

26. $\dfrac{2^{-1}p^{-2}}{q^{-3}} =$ _____

27. $\dfrac{4^{-2}a^{-3}}{b^{-4}} =$ _____

28. $4^{-3}c^{-2} =$ _____

29. $\dfrac{4^{-1}x^{-1}}{2^{-2}y} =$ _____

30. $2^{-2}5^{-2}10^2d =$ _____

31. $2^2 3^2 6^{-2}m =$ _____

32. $\dfrac{3^{-2}m}{9^{-1}n^{-1}} =$ _____

33. $\dfrac{u^{-2}v^4}{w^{-5}} =$ _____

34. $\dfrac{a^{-4}b^3}{a^{-2}b^{-3}} =$ _____

35. $\dfrac{c^{-3}d^2}{c^{-2}d^{-1}} =$ _____

36. $\dfrac{x^3y^{-4}}{3^{-2}} =$ _____

37. $\dfrac{3^{-2}b^3c^{-2}d}{2^{-3}b^{-2}c^2d^{-1}} =$ _____

38. $\dfrac{51a^{-2}b^3}{3a^{-4}b^{-3}} =$ _____

39. $\dfrac{87x^{-3}y^2}{3x^{-6}y^{-4}} =$ _____

40. $\dfrac{2^{-2}r^{-4}s^3t^{-1}}{3^{-3}r^4s^{-3}t} =$ _____

B3.5 LAWS OF EXPONENTS WITH INTEGRAL EXPONENTS

We have now defined x^n for n a positive integer, and (with $x \neq 0$) for $n = 0$, and (with $x \neq 0$, again) for n a negative integer.

Let us now summarize these definitions, along with a summary of the laws of exponents. Included among the laws will be an explicit statement of two new laws that have already been developed but not yet called laws.

> DEFINITIONS
> 1. $x^n = x \cdot x \cdot x \cdots x$ (n factors) for x any number or variable and n any positive integer.
>
> 2. $x^0 = 1$ for x any nonzero number or variable.
>
> 3. $x^{-n} = \dfrac{1}{x^n}$ for x any nonzero number or variable, and n any integer.

> LAWS OF EXPONENTS
> Let x and y be any nonzero numbers. Let m and n be any integers.
>
> First law $\qquad x^m \cdot x^n = x^{m+n}$
> Second law $\qquad (xy)^n = x^n y^n$
> Third law $\qquad (x^m)^n = x^{mn}$
> Fourth law $\qquad \left(\dfrac{x}{y}\right)^n = \dfrac{x^n}{y^n}$
> Fifth law $\qquad \dfrac{x^m}{x^n} = x^{m-n}$

Notice how often 0 is excluded. Recall why: 0^0 makes no sense and 0^n makes no sense when n is negative. But $0^n = 0$ when n is positive. In all that follows, we assume that 0 is excluded when it leads to an impossibility. In the first few examples, we will verify the common sense of the definitions and the laws for particular numerical cases.

Example A Verify the second law for $x = 3$, $y = -2$, and $n = -4$.

$$(xy)^n = [3(-2)]^{-4} = (-6)^{-4} = \frac{1}{(-6)^4} = \frac{1}{1296}$$

$$x^n y^n = 3^{-4}(-2)^{-4} = \frac{1}{3^4} \cdot \frac{1}{(-2)^4} = \frac{1}{81} \cdot \frac{1}{16} = \frac{1}{1296}$$

Example B Verify the first law for $x = -4$, $m = 3$, and $n = -5$.

$$x^m \cdot x^n = (-4)^3(-4)^{-5} = \frac{(-4)^3}{(-4)^5} = (-4)^{3-5} = (-4)^{-2} = \frac{1}{(-4)^2} = \frac{1}{16}$$

$$x^{m+n} = (-4)^{3-5} = (-4)^{-2} = \frac{1}{(-4)^2} = \frac{1}{16}$$

Example C   Verify the fifth law for $x = -3$, $m = -2$, $n = -1$.

$$\frac{x^m}{x^n} = \frac{(-3)^{-2}}{(-3)^{-1}} = \frac{(-3)^1}{(-3)^2} = (-3)^{1-2} = (-3)^{-1} = \frac{1}{(-3)^1} = -\frac{1}{3}$$

$$x^{m-n} = (-3)^{-2-(-1)} = (-3)^{-2+1} = (-3)^{-1} = \frac{1}{-3} = -\frac{1}{3}$$

In the rest of the examples, the definitions and laws will be used. The purpose of the examples and of the problems is to let you apply your knowledge of exponentiation.

Example D   Write $\left(\dfrac{2}{3}\right)^{-1}$ as a fraction in the usual sense.

$$\left(\frac{2}{3}\right)^{-1} = \frac{2^{-1}}{3^{-1}} = \frac{3}{2}$$

Example E   Write $(a^{-2}b)^2$ without using negative exponents.

$$(a^{-2}b)^2 = a^{-4}b^2 = \frac{b^2}{a^4}$$

Example F   Write $\dfrac{x^{-3}}{x^2}$ without negative exponents.

$$\frac{x^{-3}}{x^2} = x^{-3-2} = x^{-5} = \frac{1}{x^5}$$

Example G   Write $(3a^{-4})^{-2}$ without negative exponents.

$$(3a^{-4})^{-2} = 3^{-2}a^{(-4)(-2)}$$

$$= 3^{-2}a^8 = \frac{a^8}{3^2} = \frac{a^8}{9}$$

Example H   Evaluate $(2^{-3})^{-2}$.

$$(2^{-3})^{-2} = 2^6 = 64$$

Example I   Write $(3x^{-5})^{-2}$ without negative exponents.

$$(3x^{-5})^{-2} = 3^{-2}(x^{-5})^{-2}$$

$$= 3^{-2}x^{10} = \frac{x^{10}}{3^2} = \frac{x^{10}}{9}$$

B3.5 PROBLEMS

Write each expression in its usual form. Simplify where possible.

1. $\left(\dfrac{1}{2}\right)^{-3} =$ _____

2. $\left(\dfrac{3}{2}\right)^{-1} =$ _____

3. $\left(\dfrac{2}{3}\right)^{3} =$ _____

4. $\left(\dfrac{1}{3}\right)^{-2} =$ _____

5. $\left(\dfrac{4}{3}\right)^{-2} =$ _____

6. $\left(\dfrac{3}{4}\right)^{2} =$ _____

7. $(4^{-3}5^{2})^{-1} =$ _____

8. $(9^{-1}2^{3})^{2} =$ _____

9. $(3^{2}2^{-3})^{2} =$ _____

10. $(5^{-2}3^{4})^{-1} =$ _____

11. $(3^{2}9^{-1})^{-2} =$ _____

12. $(3^{5}9^{-2})^{3} =$ _____

13. $(2^{8}8^{-2})^{3} =$ _____

14. $(2^{3}8^{-1})^{-2} =$ _____

203

Write each expression without using negative exponents and simplify where possible.

15. $(x^{-1}y)^{-1} =$ _____

16. $(m^{-1}n^{-2})^{-2} =$ _____

17. $(p^{-3}q^{-2})^2 =$ _____

18. $(ab^{-1})^{-1} =$ _____

19. $\left(\dfrac{3x^{-2}}{4^{-1}}\right)^{-3} =$ _____

20. $\left(\dfrac{q^{-1}p^{-2}}{3^{-3}q^{-3}}\right)^{-3} =$ _____

21. $\left(\dfrac{8^{-1}m^{-3}}{2^{-4}n^{-2}}\right)^{-3} =$ _____

22. $\left(\dfrac{4a^{-3}}{5^{-1}}\right)^2 =$ _____

23. $(4r)^{-5} =$ _____

24. $(9x^2y^{-3})^{-2} =$ _____

25. $(5a^{-2}b^3)^{-2} =$ _____

26. $(3s^{-6})^{-4} =$ _____

27. $(x^{-1}yz^{-2})^{-2} =$ _____

28. $(u^{-1}v)^2v^{-3} =$ _____

29. $r^{-2}(rs^{-1})^3 =$ _____

30. $(b^{-2}c^{-1}d)^{-3} =$ _____

31. $(u^0v^2w^4)^{-4} =$ _____

32. $x^2x^{-5}x =$ _____

33. $(16^{-2}r^3s^{-5})^0 =$ _____

34. $(a^0b^3c^6)^{-3} =$ _____

35. $a^3aa^{-7} =$ _____

36. $\left(\dfrac{2^{-2}a^{-2}}{3^{-1}}\right)^{-3} =$ _____

37. $\left(\dfrac{3^{-2}x^{-1}}{2^{-3}}\right)^{-2} =$ _____

38. $(12^3u^{-5}v^4)^0 =$ _____

39. $\left(\dfrac{a}{b}\right)^2(a^{-1}b^{-2})^3 =$ _____

40. $(x^{-2}y^{-1})^2\left(\dfrac{x}{y}\right)^3 =$ _____

B3.6 SIMPLIFICATION OF AN EXPRESSION WITH INTEGRAL EXPONENTS

In this unit, we will use the five laws of exponents to simplify expressions. To help you, the laws of exponents are repeated.

LAWS OF EXPONENTS
Let x and y be any numbers for which a given exponential expression makes sense. Let m and n be any integers.

First Law	$x^m \cdot x^n = x^{m+n}$
Second Law	$(xy)^n = x^n y^n$
Third Law	$(x^m)^n = x^{mn}$
Fourth Law	$\left(\dfrac{x}{y}\right)^n = \dfrac{x^n}{y^n}$
Fifth Law	$\dfrac{x^m}{x^n} = x^{m-n}$

Just remember that division by 0 is not allowed and that 0^0 makes no sense.

The word **simplify** will be used in the examples and the problems. It means not to use the same base twice and to use only positive exponents. Also, where variables are used, it is assumed that they do not lead to meaningless expressions.

Example A   Simplify $(3a)^0 - 8a^0$.

Recall that unless $a = 0$, we define $a^0 = 1$. Thus,

$$(3a)^0 - 8a^0 = 1 - 8 = -7$$

Example B   Simplify $2\left(\dfrac{3}{4}\right)^{-1}$.

$$2\left(\dfrac{3}{4}\right)^{-1} = \dfrac{2 \cdot 3^{-1}}{4^{-1}} = \dfrac{2 \cdot 4}{3} = \dfrac{8}{3}$$

Example C   Simplify $2^{-1}x^3 y^{-2}$.

$$2^{-1}x^3 y^{-2} = \dfrac{x^3}{2y^2}$$

Example D   Simplify $x^{-2}(x^5 y^{-3})$.

$$x^{-2}\left(x^5 y^{-3}\right) = x^{-2+5}y^{-3} = \dfrac{x^3}{y^3}$$

Example E   Simplify $(a^{-2}b)(a^3 b^{-3})$.

$$\left(a^{-2}b\right)\left(a^3 b^{-3}\right) = a^{-2+3}b^{1-3} = a^1 b^{-2} = \dfrac{a}{b^2}$$

Example F   Simplify $p(p^4q^{-1}r^0)$.

$$p(p^4q^{-1}r^0) = p^5q^{-1} \cdot 1 = \frac{p^5}{q}$$

Example G   Simplify $\left(\dfrac{x^{-2}y}{xy^{-2}}\right)^{-1}$.

One way: $\left(\dfrac{x^{-2}y}{xy^{-2}}\right)^{-1} = \dfrac{x^2y^{-1}}{x^{-1}y^2} = \dfrac{x^{2+1}}{y^{2+1}} = \dfrac{x^3}{y^3}$

A second way: $\left(\dfrac{x^{-2}y}{xy^{-2}}\right)^{-1} = \dfrac{xy^{-2}}{x^{-2}y} = \dfrac{x^{1+2}}{y^{1+2}} = \dfrac{x^3}{y^3}$

A third way: $\left(\dfrac{x^{-2}y}{xy^{-2}}\right)^{-1} = \left(\dfrac{y^{1+2}}{x^{1+2}}\right)^{-1} = \left(\dfrac{y^3}{x^3}\right)^{-1} = \dfrac{x^3}{y^3}$

Make sure that you understand Example G and the rules involved in each of the three solutions. There are, in fact, still other ways to simplify the given expression. But this shows that there usually is no one way to go about simplifying an expression.

Example H   Simplify $a^{-3}b\left(\dfrac{b}{c}\right)^{-1}$.

$$a^{-3}b\left(\frac{b}{c}\right)^{-1} = \frac{a^{-3}bb^{-1}}{c^{-1}} = \frac{a^{-3}b^0}{c^{-1}} = \frac{c}{a^3}$$

Example I   Simplify $5^{-1} + 5^2$.

Be careful here! The answer is *not* 5.

$$5^{-1} + 5^2 = \frac{1}{5} + 25 = 25\frac{1}{5}$$

Example J   Show that $2^{-1} - 3^{-1} = (2 \cdot 3)^{-1}$.

$$2^{-1} - 3^{-1} = \frac{1}{2} - \frac{1}{3} = \frac{1}{6} = \frac{1}{2 \cdot 3} = (2 \cdot 3)^{-1}$$

Example K   Simplify $(2 - 3)^{-1}$.

$$(2 - 3)^{-1} = \frac{1}{2 - 3} = \frac{1}{-1} = -1$$

Example L   Simplify $\dfrac{8x^5y^{-2}}{2x^3yz^{-2}}$.

$$\frac{8x^5y^{-2}}{2x^3yz^{-2}} = 4x^2y^{-3}z^2 = \frac{4x^2z^2}{y^3}$$

B3.6 PROBLEMS

Simplify.

1. $4x^0 + (2x)^0 =$ _____

2. $\left(-\dfrac{3}{5}\right)^{-1} =$ _____

3. $\left(-\dfrac{2}{3}\right)^{-2} =$ _____

4. $(5m)^0 + (3m)^0 =$ _____

5. $3\left(-\dfrac{3}{5}\right)^{-2} =$ _____

6. $4^{-1}a^2b^{-3} =$ _____

7. $3^{-2}x^{-3}y =$ _____

8. $6\left(-\dfrac{2}{3}\right)^{-2} =$ _____

9. $a^{-3}(a^5b^{-3}) =$ _____

10. $2^{-3}m^{-1}n^2 =$ _____

11. $5^{-1}p^{-2}q^3 =$ _____

12. $x^4(x^{-7}y^4) =$ _____

13. $(p^{-3}q^2)(p^2q^{-3}) =$ _____

14. $(r^5s^{-2})(r^{-6}s^5) =$ _____

15. $(x^4y^{-7})(x^{-8}y^4) =$ _____

16. $(a^3b^{-4})(a^{-1}b^3) =$ _____

17. $(x^{-1}y)^2(xy^{-2})^0 =$ _____

18. $10\left(\dfrac{2}{5}\right)^{-1} =$ _____

19. $21\left(\dfrac{3}{7}\right)^{-1} =$ _____

20. $(p^{-1}q^{-2})^0(pq^{-2})^{-1} =$ _____

21. $(a^3bc^{-3})(a^{-1}b^{-1}c^{-1}) =$ _____

22. $\left(\dfrac{x^{-3}yz^{-1}}{y^{-2}z}\right)^{-1} =$ _____

23. $\left(\dfrac{p^2q^{-1}r^3}{p^{-1}r^2}\right)^{-2} =$ _____

24. $(r^{-2}s^{-1}t)(rs^{-2}t^{-1}) =$ _____

25. $\dfrac{4x^2y^{-2}z}{12xy^{-1}z^2} =$ _____

26. $2^{-1} + 3^{-1} =$ _____

27. $(2 + 3)^{-1} =$ _____

28. $\dfrac{9a^3b^{-2}c^2}{3a^{-1}b^{-1}c^{-1}} =$ _____

29. $x^{-3} + y^{-3} =$ _____

30. $(a + b)^{-2} =$ _____

31. $(x - y)^{-2} =$ _____

32. $a^{-2} + b^{-2} =$ _____

B3.7 SCIENTIFIC NOTATION

For ease of computation with very large numbers (such as astronomical distances) and with very small distances (such as weights of atomic particles), many scientists use what is called **scientific notation**.

A number is written in scientific notation if it is of the form $A \times 10^n$ where $1 \leq A < 10$ and n is an integer.

Example A   Write 345.2 in scientific notation.

$$345.2 = 3.452 \times 10^2$$

Since $1 \leq 3.452 < 10$ and 2 is an integer, this is now in scientific notation.

Example B   Write 0.03452 in scientific notation.

$$0.03452 = 3.452 \times 10^{-2}$$

Again, since $1 \leq 3.452 < 10$ and -2 is an integer, this is now in scientific notation.

Example C   Write $\frac{5}{8}$ in scientific notation.

$$\frac{5}{8} = 0.625$$
$$= 6.25 \times 10^{-1}$$

Since $1 \leq 6.25 < 10$ and -1 is an integer, $\frac{5}{8}$ has been written in scientific notation.

Example D   Write 2^{10} in scientific notation.

$$2^{10} = 1024$$
$$= 1.024 \times 10^3$$

Since $1 \leq 1.024 < 10$ and 3 is an integer, 2^{10} has been written in scientific notation.

Example E   The scientific notation for the weight of an electron is 9.107×10^{-28} grams. Write this in standard decimal notation.

The standard decimal notation for the weight is

0.0000000000000000000000000009107 grams

Example F   The speed of light is 186,282.3976 miles per second. Write this in scientific notation.

$$186,282.3976 = 1.862823976 \times 10^5$$

Example G   The U.S. national debt is in excess of two trillion dollars. Write two trillion in scientific notation.

A million is a thousand thousands:

$\qquad$ 1 million = 1,000,000

A billion is a thousand millions:

$\qquad$ 1 billion = 1,000,000,000

A trillion is a thousand billions:

$\qquad$ 1 trillion = 1,000,000,000,000

Thus,

$\qquad$ 2 trillion = 2,000,000,000,000

$\qquad\qquad\qquad = 2.0 \times 10^{12}$

This is now in scientific notation.

Scientific notation also facilitates computation.

Example H   Find the product of 80,500,000 and 0.0021 by using scientific notation.

$\qquad 80{,}500{,}000 = 8.05 \times 10^7$

and

$\qquad 0.0021 = 2.1 \times 10^{-3}$

The product is

$\qquad 8.05 \times 2.1 \times 10^7 \times 10^{-3} = 16.905 \times 10^4$

But 16.905 is not between 1 and 10, so we now write it in scientific notation. The product is

$\qquad 16.905 \times 10^4 = 1.6905 \times 10^1 \times 10^4 = 1.6905 \times 10^5$

Since $1 \le 1.6905 < 10$ and 5 is an integer, we have written the product in scientific notation.

Example I   Use scientific notation to compute $\dfrac{3804}{0.003}$ and express your answer in scientific notation.

$$\frac{3804}{0.003} = \frac{3.804 \times 10^3}{3 \times 10^{-3}} = 1.268 \times 10^6$$

B3.7 PROBLEMS

Write each number in scientific notation.

1. $365 =$ _____

2. $98.6 =$ _____

3. $5280 =$ _____

4. $1760 =$ _____

5. $105.2 =$ _____

6. $365.2422 =$ _____

7. $0.00212 =$ _____

8. $0.038 =$ _____

9. $0.0625 =$ _____

10. $0.00087 =$ _____

11. $0.000000873 =$ _____

12. $0.00000414 =$ _____

13. $1414 \times 10^{-3} =$ _____

14. $31416 \times 10^2 =$ _____

15. $271828 \times 10^5 =$ _____

16. $1732 \times 10^{-4} =$ _____

17. $\frac{3}{8} =$ _____

18. $\frac{1}{4} =$ _____

19. $\frac{9}{20} =$ _____

20. $\frac{7}{8} =$ _____

21. $\frac{1}{16} =$ _____

22. $\frac{6}{25} =$ _____

23. $\frac{367}{4} =$ _____

24. $\frac{179}{2} =$ _____

25. $3^5 =$ _____

26. $5^4 =$ _____

27. $7^4 =$ _____

28. $9^3 =$ _____

211

Write each number in standard decimal notation.

29. $314159 \times 10^{-5} =$ _____

30. $27182818 \times 10^{-7} =$ _____

31. $1732 \times 10^{-3} =$ _____

32. $1414 \times 10^{-3} =$ _____

33. $65 \times 10^4 =$ _____

34. $732 \times 10^5 =$ _____

Use scientific notation to carry out each calculation. Write each answer in scientific notation.

35. $(82 \times 10^{-3})(0.04 \times 10^5) =$ _____

36. $(0.0056 \times 10^6)(600 \times 10^{-4}) =$ _____

37. $(256 \times 10^{-5}) \div (40 \times 10^{-2}) =$ _____

38. $(243 \times 10^4) \div (900 \times 10^5) =$ _____

39. $\dfrac{(5280)(810)}{(48000)(2700)} =$ _____

40. $\dfrac{(450000)(111000)}{(3700)(600)} =$ _____

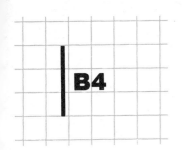

B4 OPERATIONS WITH SQUARE ROOTS

B4.1 DEFINITION OF THE RADICAL SIGN

B4.2 SIMPLIFICATION OF PRODUCTS UNDER A SINGLE RADICAL SIGN

Squaring a number (or variable) and taking the square root of a number (or variable) are opposite operations, just as addition and subtraction are opposite operations. To show that we are squaring a number, we use a superscript; for example, to square 5, we write $5 \cdot 5$, or 5^2. To square a, we write $a \cdot a$, or a^2. When squaring terms like $(a + b)$, keep in mind that $(a + b)^2$ means $(a + b)(a + b)$, which equals $a^2 + 2ab + b^2$, and not $a^2 + b^2$.

Example A Square $5x$.

$$(5x)^2 = (5x)(5x) = 5^2 x^2 = 25x^2.$$

Example B Square $(2x + y)$.

$$(2x + y)^2 = (2x + y)(2x + y)$$
$$= 4x^2 + 4xy + y^2$$

The opposite of the operation of squaring is the operation of taking the square root. The sign for this operation is "$\sqrt{}$" and is called the **radical sign**. Any numbers, variables, or expressions found inside this radical sign are to have their square roots taken. Only square roots of non-negative expressions can be taken. If A represents any non-negative expression, then $\sqrt{A}$ means a non-negative expression B for which $B^2 = A$.

Example C Square root of $49 = \sqrt{49} = 7$ because $7^2 = 49$.

Example D If $x \geq 0$, square root of $x^2 = \sqrt{x^2} = x$.

Example E   If $x \geq 0$, square root of $100x^{16} = \sqrt{100x^{16}} = 10x^8$.

In this example, we first took the square root of 100, which is 10. Then we took the square root of x^{16}. This becomes x^8 because $x^8 \cdot x^8 = x^{16}$.

Example F   If $z \geq 0$, square root of $36z^{20} = \sqrt{36z^{20}} = 6z^{10}$.

In this next example, note that we first factor the expression before taking the square root.

Example G   If $a, b \geq 0$, square root of $a^2 + 2ab + b^2 = \sqrt{a^2 + 2ab + b^2}$

$$= \sqrt{(a + b)^2}$$
$$= a + b.$$

Sometimes you can manipulate expressions under a single radical sign to make possible some simplification. This may require several steps. Even then, there are radicals that cannot be simplified.

Example H   If $a \geq 0$, simplify $\sqrt{200a^{12}}$.

Here 200 can first be factored into 100×2. 100 is a perfect square. Thus,

$$\sqrt{200a^{12}} = \sqrt{(100)(2)(a^{12})}$$

or

$$\sqrt{200a^{12}} = (\sqrt{100})(\sqrt{a^{12}})(\sqrt{2}) = 10a^6\sqrt{2}$$

Also, look for possible operations that can be performed under the radical sign before taking the square root.

Example I   If $x \geq 0$, simplify $\sqrt{18x^8 + 7x^8}$.

In this example, it is possible to first add the terms under the radical sign since they are like terms.

$$\sqrt{18x^8 + 7x^8} = \sqrt{25x^8} = 5x^4$$

Example J   If $x \geq 0$, simplify $\sqrt{(2x)(50x^5)}$.

Here we can multiply before taking the square root.

$$\sqrt{(2x)(50x^5)} = \sqrt{100x^6} = 10x^3$$

B4.1, B4.2 PROBLEMS

Simplify. Assume all variables are non-negative.

1. $\sqrt{64} =$ _____

2. $\sqrt{81} =$ _____

3. $\sqrt{x^4} =$ _____

4. $\sqrt{a^8} =$ _____

5. $\sqrt{w^{32}} =$ _____

6. $\sqrt{25x^2} =$ _____

7. $\sqrt{16x^{14}} =$ _____

8. $\sqrt{144b^8} =$ _____

9. $\sqrt{100a^{10}} =$ _____

10. $\sqrt{16x^6} =$ _____

11. $\sqrt{75x^{12}} =$ _____

12. $\sqrt{50x^4} =$ _____

13. $\sqrt{20a^{18}} =$ _____

14. $\sqrt{18x^6} =$ _____

15. $\sqrt{27a^2} =$ _____

16. $\sqrt{32q^{16}} =$ _____

17. $\sqrt{300a^{30}} =$ _____

18. $\sqrt{400w^{10}} =$ _____

19. $\sqrt{12p^{24}} =$ _____

20. $\sqrt{25x^{50}} =$ _____

219

21. $\sqrt{2b^2 + 2b^2} =$ _____

22. $\sqrt{3a^2 + 6a^2} =$ _____

23. $\sqrt{x^6 + 8x^6} =$ _____

24. $\sqrt{5a^4 + 20a^4} =$ _____

25. $\sqrt{12x^8 + 13x^8} =$ _____

26. $\sqrt{(8x^2)(2x^4)} =$ _____

27. $\sqrt{(2a^5)(18a)} =$ _____

28. $\sqrt{5x^2 - x^2} =$ _____

29. $\sqrt{(18w^4)(2x^4)} =$ _____

30. $\sqrt{(5z^5)(5z)} =$ _____

31. $\sqrt{\dfrac{9}{16}} =$ _____

32. $\sqrt{\dfrac{25}{81}} =$ _____

33. $\sqrt{(9a^3)(4a^3)} =$ _____

34. $\sqrt{12x^4 - 3x^4} =$ _____

35. $\sqrt{(2x^2)(2x^2)} =$ _____

36. $\sqrt{(3x^6)(12x^6)} =$ _____

37. $\sqrt{9b^2 + 16b^2} =$ _____

38. $\sqrt{18a^2b^2} =$ _____

39. $\sqrt{(a + b)^2} =$ _____

40. $\sqrt{x^2 + 2xy + y^2} =$ _____

41. $\sqrt{27a^4b^2} =$ _____

42. $\sqrt{16x^2y^4z^6} =$ _____

43. $\sqrt{\dfrac{36a^2b^2}{49}} =$ _____

44. $\sqrt{\dfrac{25y^3}{64}} =$ _____

B4.3 ADDITION AND SUBTRACTION OF RADICAL EXPRESSIONS

B4.4 MULTIPLICATION AND DIVISION OF RADICAL EXPRESSIONS

Performing addition, subtraction, multiplication, and division with radical expressions requires very careful attention. For example,

$$\sqrt{4} \cdot \sqrt{9} = \sqrt{36} = 6 \quad \text{however} \quad \sqrt{4} + \sqrt{9} \neq \sqrt{13}$$

We will study multiplication and division first. In general, when multiplying radical expressions, we simply multiply the terms and put them under one radical sign. Notice in the first example why this is possible.

Example A   Method 1: Take the square root before multiplying.

$$\sqrt{4} \cdot \sqrt{9} = 2 \times 3 = 6$$

Method 2: Multiply under one radical sign and take the square root.

$$\sqrt{4} \cdot \sqrt{9} = \sqrt{36} = 6$$

Example B   Find $\sqrt{2}\sqrt{32}$.

In this situation, method 2 is easier.

$$\sqrt{2}\sqrt{32} = \sqrt{64} = 8$$

Notice, however, that method 1 will work.

$$\sqrt{2}\sqrt{16 \cdot 2} = \sqrt{2} \cdot 4\sqrt{2} = 4 \cdot 2 = 8$$

Example C   For $a, b \geq 0, \sqrt{3ab}\sqrt{12ab} = \sqrt{36a^2b^2} = 6ab$.

Example D   For $x \geq 0, \sqrt{2x^3}\sqrt{8x^5} = \sqrt{16x^8} = 4x^4$.

The rule for dividing radical expressions can be illustrated by these examples.

Example E   Find $\dfrac{\sqrt{16}}{\sqrt{4}}$.

Method 1 $\dfrac{\sqrt{16}}{\sqrt{4}} = \dfrac{4}{2} = 2$.   Method 2 $\dfrac{\sqrt{16}}{\sqrt{4}} = \sqrt{\dfrac{16}{4}} = \sqrt{4} = 2$.

In general, for $x \geq 0, y > 0, \dfrac{\sqrt{x}}{\sqrt{y}} = \sqrt{\dfrac{x}{y}}$.

Example F   For $a, b > 0$, $\dfrac{\sqrt{72a^3b^5}}{\sqrt{2ab}} = \sqrt{\dfrac{72a^3b^5}{2ab}} = \sqrt{36a^2b^4} = 6ab^2$.

For adding and subtracting radical expressions, there are no general rules to assist us. The best we can hope for is that the expressions may have factors that make it possible to collect like terms or to use the distributive property.

Example G $\sqrt{5} + \sqrt{5} = \sqrt{5}(1 + 1) = 2\sqrt{5}$.

$4\sqrt{6} - \sqrt{6} = \sqrt{6}(4 - 1) = 3\sqrt{6}$.

Example H $\sqrt{36} + \sqrt{16} = 6 + 4 = 10$, and $\sqrt{9} - \sqrt{4} = 3 - 2 = 1$.

However, the trap that you must constantly look out for is best illustrated in the two examples below. Study them carefully.

Example I $\sqrt{9} + \sqrt{16} = 3 + 4 = 7$, whereas $\sqrt{9 + 16} = \sqrt{25} = 5$.

The same holds true for subtraction.

Example J $\sqrt{25} - \sqrt{9} = 5 - 3 = 2$, whereas $\sqrt{25 - 9} = \sqrt{16} = 4$.

There are times when it is possible to perform addition and subtraction on radical expressions. This often involves factoring and the distributive property.

Example K $\sqrt{8} + \sqrt{50} = \sqrt{2 \cdot 2 \cdot 2} + \sqrt{2 \cdot 5 \cdot 5}$

$= 2\sqrt{2} + 5\sqrt{2} = \sqrt{2}(2 + 5) = 7\sqrt{2}$

Example L $\sqrt{9x} + \sqrt{x^3} = 3\sqrt{x} + x\sqrt{x} = \sqrt{x}(3 + x)$.

Example M $\sqrt{75} - \sqrt{27} = \sqrt{5 \times 5 \times 3} - \sqrt{3 \times 3 \times 3}$

$= 5\sqrt{3} - 3\sqrt{3} = \sqrt{3}(5 - 3) = 2\sqrt{3}$

Example N   For $x \geq 0$, $\sqrt{25x^3} - \sqrt{4x} = 5x\sqrt{x} - 2\sqrt{x} = \sqrt{x}(5x - 2)$.

In general, it pays to express radicals in their simplest form before beginning any operation(s) with them.

Example O   For $x \geq 0$, $\sqrt{5x}(\sqrt{5} + \sqrt{x}) = \sqrt{25x} + \sqrt{5x^2} = 5\sqrt{x} + x\sqrt{5}$.

In many situations, the form of the final answer is sometimes arbitrary. Often it depends upon what use you will be making of the result.

B4.3, B4.4 PROBLEMS

Perform the indicated operations. Presume that all variables are positive.

1. $(\sqrt{6})(\sqrt{6}) = $ _____

2. $(\sqrt{8})(\sqrt{2}) = $ _____

3. $(\sqrt{16})(\sqrt{36}) = $ _____

4. $(\sqrt{2})(\sqrt{18}) = $ _____

5. $(\sqrt{32})(\sqrt{2}) = $ _____

6. $(\sqrt{9a^2})(\sqrt{4a^2}) = $ _____

7. $(\sqrt{25x})(\sqrt{4x}) = $ _____

8. $(\sqrt{3x^3})(\sqrt{12x^7}) = $ _____

9. $(\sqrt{x} - 2)^2 = $ _____

10. $(\sqrt{3})(\sqrt{3x^8}) = $ _____

11. $(\sqrt{b})(\sqrt{b} + \sqrt{4b^3}) = $ _____

12. $(\sqrt{5w})(\sqrt{5} + \sqrt{w}) = $ _____

13. $a(\sqrt{a^3} - \sqrt{a}) = $ _____

14. $\sqrt{b}(\sqrt{a} + \sqrt{b}) = $ _____

15. $(\sqrt{2} + \sqrt{32}) = $ _____

16. $\dfrac{\sqrt{4}}{\sqrt{9}} = $ _____

17. $\dfrac{\sqrt{ab^2}}{\sqrt{a}} = $ _____

18. $\dfrac{\sqrt{x^2 y^4}}{xy} = $ _____

223

19. $\dfrac{ab}{\sqrt{a^2b^2}} =$ _____

20. $\dfrac{a + 2}{\sqrt{a^2} + \sqrt{4}} =$ _____

21. $\sqrt{16} + \sqrt{9} =$ _____

22. $\sqrt{25} - \sqrt{4} =$ _____

23. $\sqrt{36} + \sqrt{9} =$ _____

24. $\sqrt{12} - 2\sqrt{3} =$ _____

25. $\sqrt{8} + \sqrt{32} =$ _____

26. $\sqrt{50} - \sqrt{18} =$ _____

27. $\sqrt{9} + \sqrt{121} =$ _____

28. $\sqrt{18} - \sqrt{9} =$ _____

29. $\sqrt{4a} + \sqrt{a^3} =$ _____

30. $\sqrt{x^4} - \sqrt{x^5} =$ _____

31. $\sqrt{b^3} + \sqrt{4b^2} =$ _____

32. $\sqrt{25a^2} - \sqrt{a^4} =$ _____

33. $\sqrt{9x^3} + \sqrt{36x} =$ _____

34. $\sqrt{2a^2} + \sqrt{2b^2} =$ _____

35. $\sqrt{5x^2} - \sqrt{3x^3} =$ _____

36. $\sqrt{8b^2} + \sqrt{28b^2} =$ _____

37. $\sqrt{25x} + \sqrt{x^3} =$ _____

38. $\sqrt{4w} - \sqrt{4z} =$ _____

39. $\sqrt{9x^3} - \sqrt{9x^2y} =$ _____

40. $\sqrt{a^4} + \sqrt{b^4} =$ _____

B4.5 SOLUTION OF A SIMPLE RADICAL EQUATION

Radicals frequently show up in equations. In general, they do not present any unusual problems. The rules for order of operations should be followed.

Example A   Solve for x.

$$\sqrt{x} + 3 = 8$$

You can ask, "What must be added to 3 to equal 8?" The answer is 5.

$$\sqrt{x} = 5$$
$$x = 25$$

Example B   Solve for a.

$$\sqrt{a} - 2 = 7$$
$$\sqrt{a} = 9$$
$$a = 81$$

Example C   Solve for z.

$$\sqrt{25z} + 4 = 19$$

Here you have two choices to begin with. Either subtract 4 from both sides, or take the square root of 25. It does not matter which is done first.

$$\sqrt{25z} + 4 - 4 = 19 - 4 \quad \text{Subtract 4}$$
$$\sqrt{25z} = 15$$
$$5\sqrt{z} = 15 \quad \text{Take the square root of 25}$$
$$\frac{5\sqrt{z}}{5} = \frac{15}{5} \quad \text{Divide both sides by 5}$$
$$\sqrt{z} = 3$$
$$z = 9$$

Example D   Solve for m.

$$\frac{\sqrt{m}}{4} = 1$$
$$\frac{4\sqrt{m}}{4} = (4)(1) \quad \text{Multiply both sides by 4}$$
$$\sqrt{m} = 4$$
$$m = 16$$

Example E   Solve for x.

$$2\sqrt{x} - 3 = 5$$

$$2\sqrt{x} - 3 + 3 = 5 + 3 \quad \text{Add 3 to both sides}$$

$$2\sqrt{x} = 8$$

$$\frac{2\sqrt{x}}{2} = \frac{8}{2} \quad \text{Divide both sides by 2}$$

$$\sqrt{x} = 4$$

$$x = 16 \quad \text{Square both sides}$$

Example F   Solve for x.

$$\left(2\sqrt{x}\right)^2 + 3 = 11$$

$$\left(2\sqrt{x}\right)^2 + 3 - 3 = 11 - 3 \quad \text{Subtract 3}$$

$$\left(2\sqrt{x}\right)^2 = 8$$

$$4x = 8 \quad \text{Square the } \left(2\sqrt{x}\right) \text{ term}$$

$$x = 2 \quad \text{Divide by 4}$$

To solve some equations, a series of steps may have to be taken before a solution becomes apparent.

Example G   Solve for x.

$$\sqrt{9x} + \sqrt{x} = 8$$

$$3\sqrt{x} + \sqrt{x} = 8 \quad \text{Take the square root of 9}$$

$$4\sqrt{x} = 8 \quad \text{Add}$$

$$\sqrt{x} = 2 \quad \text{Divide}$$

$$x = 4 \quad \text{Square both sides of the equation}$$

This last example illustrates how an equation that seems unsolvable at first glance can be solved with some manipulation.

B4.5 PROBLEMS

Solve each equation for the variable, which is assumed to be greater than or equal to 0.

1. $\sqrt{a} + 1 = 3$ _____

2. $\sqrt{4} + \sqrt{x} = 9$ _____

3. $5 + \sqrt{a^2} = 7$ _____

4. $6 + \sqrt{2x} = 8$ _____

5. $8 - \sqrt{x} = 3$ _____

6. $\sqrt{2x} - 6 = 0$ _____

7. $\sqrt{x} + \sqrt{x} = 4$ _____

8. $\sqrt{y} - 3 = 6$ _____

9. $\sqrt{3x} - 2 = 1$ _____

10. $\sqrt{5w} - 2 = 3$ _____

11. $\sqrt{10x} - 8 = 2$ _____

12. $\sqrt{6a} + \sqrt{6a} = 12$ _____

13. $3\sqrt{x} - 1 = 5$ _____

14. $4\sqrt{x} - 1 = 7$ _____

227

15. $12 - \sqrt{2x} = 8$ _____

16. $9 - 3\sqrt{x} = 3$ _____

17. $3\sqrt{x} + 2\sqrt{x} = 10$ _____

18. $5\sqrt{a} - 2\sqrt{a} = 9$ _____

19. $2(\sqrt{b} + 1) = 6$ _____

20. $3(\sqrt{x} + 1) = 6$ _____

21. $\dfrac{\sqrt{x}}{4} = 1$ _____

22. $\dfrac{3\sqrt{a}}{5} = 3$ _____

23. $\dfrac{\sqrt{x}}{18} = 2$ _____

24. $5\sqrt{z} + 1 = 1$ _____

25. $3\sqrt{k} + 6 = 9$ _____

26. $2\sqrt{q} - 1 = 5$ _____

27. $\left(\sqrt{p}\right)^2 = 36$ _____

28. $\dfrac{\sqrt{q}}{2} = 18$ _____

29. $2(\sqrt{x} + 3) = 6$ _____

30. $\left(2\sqrt{s}\right)^2 + 3 = 5$ _____

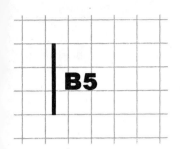

B5

SOLUTION OF LINEAR EQUATIONS AND INEQUALITIES

B5.1 SOLUTION OF A LINEAR EQUATION IN ONE UNKNOWN WITH NUMERICAL COEFFICIENTS

In mathematics, the skill of **solving equations** is a very important and useful one. You did some of this in B4.5. In this unit, you will solve **linear equations**.

An equation is linear if it can be written in the form

(Number)(Variable) + (Number) = 0

For example, $7x - 3 = 0$ is in this form.

Example A   Is $12x - 4 = 8x - 5$ a linear equation?

The answer is yes. Here's why.

$12x - 4 = 8x - 5$

Subtract $8x$ from each side

$4x - 4 = -5$ Add 5 to each side

$4x + 1 = 0$

This is now in the form

(Number)(Variable) + (Number) = 0

so it is a linear equation.

Example B   Solve $12x - 4 = 8x - 5$.

We've seen that $12x - 4 = 8x - 5$ can be rewritten as $4x + 1 = 0$. To solve this, all you have to do is to isolate x. To this end, subtract 1 from both sides to obtain $4x = -1$. Now divide both sides by 4 to get $x = -\dfrac{1}{4}$.   **229**

This is the solution. Check that it makes the original equation true by substituting $-\frac{1}{4}$ for x in each side of $12x - 4 = 8x - 5$.

Example C   Solve $3(2x - 4) = 6$.

As before, simplify and isolate the variable x.

$$3(2x - 4) = 6 \qquad \text{Clear the parentheses}$$
$$6x - 12 = 6 \qquad \text{Add 12 to each side}$$
$$6x = 18 \qquad \text{Divide both sides by 6}$$
$$x = 3 \qquad \text{This is the solution}$$

Example D   If $3(2t + 3) - 4(3t - 2) = 47$, then what must t equal?

$$3(2t + 3) - 4(3t - 2) = 47 \qquad \text{Clear parentheses}$$
$$6t + 9 - 12t + 8 = 47 \qquad \text{Collect like terms}$$
$$-6t + 17 = 47 \qquad \text{Subtract 17}$$
$$-6t = 30 \qquad \text{Divide by } -6$$
$$t = -5 \qquad \text{This is the solution}$$

Example E   If I had \$3 less than I do, I would have $\frac{4}{5}$ as much money as I do. How much money do I have?

This is an example of a *word problem*. To solve it, first rewrite the problem as an algebra problem. To that end, let $M =$ amount of money I have. (Try always to use a suggestive letter for the variable.) Then \$3 less is $M - 3$. Thus,

$$M = \text{Amount of money I have}$$

$$M - 3 = \$3 \text{ less}$$

$$\frac{4}{5}M = \frac{4}{5} \text{ as much money as I have}$$

The problem says that

$$M - 3 = \frac{4}{5}M$$

and this is the equation to solve.

$$M - 3 = \frac{4}{5}M \qquad \text{Multiply by 5}$$
$$5M - 15 = 4M \qquad \text{Subtract } 4M$$
$$M - 15 = 0 \qquad \text{Add 15}$$
$$M = 15$$

I have \$15.

B5.1 PROBLEMS

Show that each of the following is a linear equation (see Example A for a reminder of how to do it).

1. $4x - 8 = x - 11$

2. $6t + 7 = 2t - 3$

3. $9 - 5s = 7s + 6$

4. $6y - 8 = 5 - 2y$

5. $3(n - 4) + 2n = 5(1 - n)$

6. $2(3 + m) = 4 - 3(m - 1)$

7. $x(y - 2) - 3 = (3 + y)x$

8. $s(4 - t) = 5 - s(t + 2)$

Solve each linear equation for the variable indicated.

9. $2x + 3 = 9$

$x = $ _____

10. $5t - 4 = 6$

$t = $ _____

11. $-12 - 4s = 8$

$s = $ _____

12. $15 + 10r = 5$

$r = $ _____

13. $7n - 5 = 4n + 3$

$n = $ _____

14. $3(s - 7) = 1 - 2s$

$s = $ _____

15. $4m - 5 = 3(2 - m)$

$m = $ _____

16. $6k + 1 = 5 - 3k$

$k = $ _____

17. $\frac{2}{3}x + 4 = \frac{3}{2} - \frac{5}{3}x$

$x = $ _____

18. $6 - \frac{4}{5}y = \frac{1}{2}y + \frac{3}{5}$

$y = $ _____

19. $\frac{3s - 2}{4} = 5 - s$

$s = $ _____

20. $3(2n - 3) - (2 - n) = 4$

$n = $ _____

21. $2(4 - k) - (k - 3) = -1$

$k = $ _____

22. $x + 7 = \frac{4 - 5x}{2}$

$x = $ _____

23. $6 - 3x = \frac{2x - 3}{5} + 1$

$x = $ _____

24. $\frac{2 - 6s}{5} = \frac{4s - 3}{2}$

$s = $ _____

25. $\frac{3r - 2}{4} = \frac{3 - 4r}{3}$

$r = $ _____

26. $\frac{4y - 7}{5} = 2 - (4 - y)$

$y = $ _____

Solve.

27. Five years from now, Sam will be twice as old as he was 8 years ago. How old is Sam now? (*Hint*: Let S be Sam's age now. In 5 years, he will be $S + 5$ years old, and he was $S - 8$ years old 8 years ago.)

28. There are 30 pieces of fruit in a bowl of apples, oranges, and pears. There are twice as many oranges as apples, and half as many apples as pears. How many of each type are there? (*Hint*: Let A = number of apples. Now express the numbers of oranges and pears in terms of A. They must add up to 30.)

29. Twenty less than five times a number is ten. Find the number.

30. Six more than three-fourths of a number is eighteen. Find the number.

B5.2 SOLUTION OF A LINEAR EQUATION IN ONE UNKNOWN WITH LITERAL COEFFICIENTS

As you have seen, letting letters stand for numbers is one of the main ideas of beginning algebra. Thus, if $2x + 3 = 9$, we can solve for x.

$2x + 3 = 9$	Subtract 3
$2x = 6$	Divide by 2
$x = 3$	3 is the solution

In like manner, we can solve $2x + a = 9$ for the unknown x.

$2x + a = 9$	Subtract a
$2x = 9 - a$	Divide by 2
$x = \dfrac{9 - a}{2}$	$\dfrac{9 - a}{2}$ is the solution

Example A   Solve $3ax + 2 = 5$ for x.

$3ax + 2 = 5$	Subtract 2
$3ax = 3$	Divide by $3a$
$x = \dfrac{3}{3a} = \dfrac{1}{a}$	$\dfrac{1}{a}$ is the solution

Example B   Solve $2bt - 3b = 1$ for t.

$2bt - 3b = 1$	Add $3b$
$2bt = 1 + 3b$	Divide by $2b$
$t = \dfrac{1 + 3b}{2b}$	$\dfrac{1 + 3b}{2b}$ is the solution

The equations in Examples A and B are said to have *literal* (letter) coefficients because the multipliers or constant terms, or both, involve letters rather than just numbers.

Example C   Solve $5F - 9C = 160$ for F.

$5F - 9C = 160$	Add $9C$
$5F = 9C + 160$	Divide by 5
$F = \dfrac{9}{5}C + 32$	

This last equation is the formula for converting temperatures from Celsius to Fahrenheit.

Example D Solve $5F - 9C = 160$ for C.

$$5F - 9C = 160 \qquad \text{Subtract } 5F$$

$$-9C = -5F + 160 \qquad \text{Multiply by } -1$$

$$9C = 5F - 160 \qquad \text{Factor out 5 on the right}$$

$$9C = 5(F - 32) \qquad \text{Divide by 9}$$

$$C = \frac{5}{9}(F - 32)$$

This time, we get another temperature conversion formula from Fahrenheit to Celsius.

Example E Solve $A = h \cdot \dfrac{B + b}{2}$ for B. (Area of a trapezoid)

$$A = h \cdot \frac{B + b}{2} \qquad \text{Multiply by 2}$$

$$2A = h(B + b) \qquad \text{Clear the parentheses}$$

$$2A = hB + hb \qquad \text{Subtract } hb$$

$$2A - hb = hB \qquad \text{Divide by } h$$

$$B = \frac{2A - hb}{h}$$

Example F Solve $V = \dfrac{1}{3}\pi r^2 h$ for h. (Volume of a cone)

$$V = \frac{1}{3}\pi r^2 h \qquad \text{Multiply by 3}$$

$$3V = \pi r^2 h \qquad \text{Divide by } \pi r^2$$

$$h = \frac{3V}{\pi r^2}$$

Example G Solve $I = Prt$ for r. (Interest formula)

$$I = Prt \qquad \text{Divide by } Pt$$

$$r = \frac{I}{Pt}$$

Example H $R = P(1 + rt)$ for r. (Interest formula)

$$R = P(1 + rt) \qquad \text{Clear the parentheses}$$

$$R = P + Prt \qquad \text{Subtract } P$$

$$R - P = Prt \qquad \text{Divide by } Pt$$

$$r = \frac{R - P}{Pt}$$

B5.2 PROBLEMS

Solve each linear equation for the designated variable. Simplify your answer where possible.

1. $5ax - 2 = 3$, for $x = $ _____

2. $4 - 6cy = 10$, for $y = $ _____

3. $7 - 4bt = 3$, for $t = $ _____

4. $8cs + 3 = 19$, for $s = $ _____

5. $3ay - 5b = 4$, for $y = $ _____

6. $2A - 3Ax = 5A$, for $x = $ _____

7. $6d + 5ds = -4d$, for $s = $ _____

8. $4bx - 7c = 2$, for $x = $ _____

9. $F = 2C + 30$, for $C = $ _____

10. $C = \frac{1}{2}(F - 30)$, for $F = $ _____

11. $S = 2\pi r^2 + 2\pi rh$, for $h = $ _____

12. $R = P + Prt$, for $t = $ _____

13. $R = P + Prt$, for $P = $ _____

14. $S = \pi r^2 + \frac{1}{3}\pi r^2 h$, for $h = $ _____

235

15. $15ct - 8t - 9 = 0$, for $t =$ _____

16. $3ab + 4bc = 3$, for $b =$ _____

17. $9xy - 3yz = 2$, for $y =$ _____

18. $9rx + 6x + 5 = 0$, for $x =$ _____

19. $(3b - 2)r + 3r = b$, for $r =$ _____

20. $ab - 2bc + 3ac = 0$, for $a =$ _____

21. $ab - 2bc + 3ac = 0$, for $b =$ _____

22. $(5 - 2a)s - 4s = 12a$, for $s =$ _____

23. $\dfrac{3ax - b}{2} = 4ax + 3x$, for $x =$ _____

24. $2ay(a - 3) - (2 - 4a)y = a^2$, for $y =$ _____

25. $(3c - 2)s + 3cs(2c - 1) = c + 2$, for $s =$ _____

26. $\dfrac{5mn - 3n}{4} = m + 2n$, for $n =$ _____

Solve.

27. I'm thinking of a number. If I increase it by an amount A, then I get double the number. What is the number?

28. Jay's boss increased Jay's salary by half. This caused Jay to make D more than he was making. How much was he making?

B5.3 RATIO AND PROPORTION

A **ratio** is a way of relating a pair of numbers or of quantities that are not necessarily in the same units. A ratio usually describes a comparison or a rate.

Example A In Professor Messer's class, there are 14 males and 16 females. What is the ratio of males to females?

The ratio of males to females is "14 to 16." This is written either as $14:16$ or $\frac{14}{16}$ (but be careful; not all ratios are fractions—indeed, this one is not, since the first member counts males while the second counts females).

Example B Can the ratio in Example A be expressed in another way?

Yes. The ratio of "14 to 16" is equivalent to the ratio of "7 to 8." Though ratios are not necessarily fractions, equality of ratios is just like equality of fractions, and $\frac{14}{16} = \frac{7}{8}$. If two ratios are equal, they form a **proportion**.

Example C Lobster tails are on sale at 9 for $6. What is the cost of a dozen lobster tails?

Let C be the cost of a dozen lobster tails. Then we want this proportion:

$$\frac{9}{6} = \frac{12}{C}$$ Multiply through by $6C$

$$9C = 6 \cdot 12 = 72$$ Divide through by 9

$$C = 8$$

A dozen lobster tails will cost $8.

Example D If a car travels M miles on 8 gallons of gasoline, how far will it travel on 5 gallons?

We set up the proportion

$$\frac{M}{8} = \frac{d}{5}$$

where d is the distance it will travel on 5 gallons. Now solve for d:

$$\frac{M}{8} = \frac{d}{5}$$ Multiply by 40

$$5M = 8d$$ Divide by 8

$$d = \frac{5M}{8}$$

The car will travel $\frac{5M}{8}$ miles on 5 gallons of gasoline.

In Example D, the quantities M and d are **directly proportional**. In general, the quantity y is called proportional (or directly proportional) to the quantity x if there is a constant K for which $y = Kx$. K is called the **constant of proportionality**.

Example E Ed drives 55 miles per hour. How long will it take him to go 495 miles?

Here the distance d and the time t are directly proportional:

$$(\text{Distance}) = (\text{Speed}) \cdot (\text{Time})$$

Since $d = 495$, we have

$$495 = 55t$$

$$t = \frac{495}{55} = 9$$

It will take Ed 9 hours.

In Example E, the quantities t and s are **inversely proportional**. In general, y is inversely proportional to x if there is a constant K for which $y = \dfrac{K}{x}$. Note that $x = \dfrac{K}{y}$. K is called the **constant of inverse proportionality**.

Example F P is directly proportional to Q, and when $P = 7$, $Q = 13$. Find the constant of proportionality K, and tell what P is when $Q = 91$.

First, we have

$$P = KQ$$

Set $P = 7$ and $Q = 13$.

$$7 = 13K$$

$$K = \frac{7}{13}$$

Thus,

$$P = \frac{7}{13} Q$$

Now set $Q = 91$.

$$P = \frac{7}{13} \cdot 91 = 7 \cdot 7 = 49$$

When $Q = 91$, $P = 49$.

B5.3 PROBLEMS

Solve. Write each ratio in the form $\frac{a}{b}$.

1. In Gus's restaurant, there are 3 waiters for every 7 tables. What is the ratio of waiters to tables?

2. Give the ratio of vowels (a, e, i, o, u) to consonants in the word *facetious*.

3. Give the ratio of consonants to vowels (a, e, i, o, u) in the word *abstemious*.

4. In the month of January, what is the ratio of even dates to odd dates?

5. In Example A, what is the ratio of females to males?

6. In Example A, what is the ratio of males to students?

7. Do the ratios $\frac{54}{36}$ and $\frac{57}{38}$ form a proportion? Explain.

8. Do the ratios $\frac{111}{37}$ and $\frac{3}{1}$ form a proportion? Explain.

9. Find x so that the ratios $\frac{x}{21}$ and $\frac{28}{12}$ form a proportion.

10. Find x so that the ratios $\frac{48}{64}$ and $\frac{57}{x}$ form a proportion.

11. The HMS *Berlin* traveled 745.2 nautical miles in 3 days. At this rate, how far will be traveled in 4 days?

12. In the human body, blood travels at the rate of about 4.8 meters in 5 seconds. How far will it travel in 2 minutes?

13. A local auto dealer keeps an inventory of sedans and coupes in the ratio of 4 : 7. He feels he can sell 33 sedans next month. About how many coupes should he have on hand?

14. A steamship keeps a passenger-to-crew ratio of 20 : 13. On the next trip, they will have 1423 passengers. About how many crew members should be aboard?

15. The human heart pumps 7.5 quarts of blood in 3 minutes. How many does it pump in one day?

16. The human heart beats 324 times in 4 minutes. How many times does it beat in a day?

17. A theatre counts on selling 9 cups of soda for every 4 boxes of popcorn. Last night, they sold 135 boxes of popcorn. About how many cups of soda did they sell?

18. Last year on a 3-day vacation weekend, there were 521 accidental deaths on the highways. This year, that weekend has been extended to 4 days. About how many highway deaths can be expected this year?

19. A restaurant normally sells 8 roast beef dinners for every 5 steaks it sells. Tonight, they expect to sell 63 steaks. About how many roast beef dinners can they expect to sell?

20. A textbook publisher last year sold $972,358 worth of books from their list of 37 titles. If books sell at the same rate as last year, about how many dollars worth of books can the publisher expect to sell this year on a list of 43 titles?

The following problems concern direct and inverse proportions.

21. A is directly proportional to B. The constant of proportionality is 3.5. What is B when A is 21?

22. M is inversely proportional to N, and the constant of inverse proportionality is 15. What is N when $M = 5$?

23. For automobile gears, the number of teeth and the number of revolutions vary inversely. A gear with 42 teeth meshes with another that has 56 teeth. When the first gear makes 3500 revolutions, how many revolutions does the second gear make?

24. For a savings account, the yearly interest is directly proportional to the amount of money in the account. If the yearly interest on $1800 is $126, what is the yearly interest on $2200?

25. S is inversely proportional to T. When $S = 9$, $T = 2$. If $S = 6$, what is T?

26. X is inversely proportional to Y, and $X = 9$ when $Y = 4$. What is X when $Y = 6$?

B5.4 SOLUTION OF A SIMPLE EQUATION IN ONE UNKNOWN THAT IS REDUCIBLE TO A LINEAR EQUATION

Sometimes, equations don't *look* linear, but they can be transformed into linear equations.

Example A Solve $\dfrac{2x + 4}{x} = 3$.

This doesn't look like a linear equation, and indeed it isn't, but just multiply both sides by x: then you get to a linear equation. Here's how it goes:

$$\frac{2x + 4}{x} = 3 \qquad \text{Multiply by } x$$

$$2x + 4 = 3x \qquad \text{Subtract } 2x$$

$$4 = x \qquad \text{The solution is 4}$$

Check the solution:

$$\frac{2 \cdot 4 + 4}{4} = \frac{12}{4} = 3$$

Example B Solve $\dfrac{1}{3} + \dfrac{1}{x} = \dfrac{1}{2}$.

$$\frac{1}{3} + \frac{1}{x} = \frac{1}{2} \qquad \text{Multiply by } 6x$$

$$2x + 6 = 3x \qquad \text{Subtract } 2x$$

$$6 = x$$

Check: $\dfrac{1}{3} + \dfrac{1}{6} = \dfrac{2}{6} + \dfrac{1}{6} = \dfrac{3}{6} = \dfrac{1}{2}$

Example C Solve $\dfrac{2x + 3}{3x - 2} = 5$.

$$\frac{2x + 3}{3x - 2} = 5 \qquad\qquad \text{Multiply by } 3x - 2$$

$$2x + 3 = 5(3x - 2) \qquad \text{Clear the parentheses}$$

$$2x + 3 = 15x - 10 \qquad \text{Subtract } 2x$$

$$3 = 13x - 10 \qquad \text{Add 10}$$

$$13x = 13 \qquad\qquad \text{Divide by 13}$$

$$x = 1$$

Check: $\dfrac{2 \cdot 1 + 3}{3 \cdot 1 - 2} = \dfrac{2 + 3}{3 - 2} = \dfrac{5}{1} = 5$

Example D   Solve $\dfrac{2}{t} - 3 = 4 - \dfrac{3}{t}$.

$$\dfrac{2}{t} - 3 = 4 - \dfrac{3}{t} \qquad \text{Multiply by } t$$

$$2 - 3t = 4t - 3 \qquad \text{Add } 3t$$

$$2 = 7t - 3 \qquad \text{Add } 3$$

$$5 = 7t \qquad \text{Divide by } 7$$

$$t = \dfrac{5}{7}$$

Example E   Solve $\dfrac{6}{s - 4} + \dfrac{4}{s - 6} = \dfrac{2}{s - 4}$.

Here we must multiply by a factor that will clear all the denominators. A simple glance at the denominators indicates that multiplying by $(s - 4)(s - 6)$ will work.

$$\dfrac{6}{s - 4} + \dfrac{4}{s - 6} = \dfrac{2}{s - 4} \qquad \text{Multiply by } (s - 4)(s - 6)$$

$$6(s - 6) + 4(s - 4) = 2(s - 6) \qquad \text{Clear the parentheses}$$

$$6s - 36 + 4s - 16 = 2s - 12 \qquad \text{Add } 36$$

$$6s + 4s - 16 = 2s + 24 \qquad \text{Add } 16$$

$$6s + 4s = 2s + 40 \qquad \text{Subtract } 2s$$

$$6s + 4s - 2s = 40 \qquad \text{Add like terms}$$

$$8s = 40 \qquad \text{Divide by } 8$$

$$s = 5$$

Let us check to make sure that this is indeed a solution to the original equation.

$$\dfrac{6}{5 - 4} + \dfrac{4}{5 - 6} = 6 - 4 = 2$$

$$\dfrac{2}{5 - 4} = 2$$

Thus, $\dfrac{6}{s - 4} + \dfrac{4}{s - 6} = \dfrac{2}{s - 4}$ when $s = 5$.

The importance of checking the solution in the original equation has to do with the fact that some "solutions" aren't really solutions. Example F will be a case in point.

Example F   Solve $\dfrac{2}{t-1} - \dfrac{1}{2t-2} = \dfrac{3}{2t-2} - 2$.

Multiply through by $2t - 2$, and you get

$$4 - 1 = 3 - 4t + 4$$

$$3 = 7 - 4t$$

$$4t = 4$$

$$t = 1$$

Now when you substitute $t = 1$ in the original equation, you get

$$\frac{2}{0} - \frac{1}{0} = \frac{3}{0} - 2$$

But in a fraction, the denominator can *never* be 0! Thus, the "solution" $t = 1$ is not a solution. In fact, the original equation has no solution.

Example G   Solve $x(5 - 2x) = 2(3 - x^2)$.

This does not look like a linear equation because it contains terms involving x^2. But take a closer look.

$$x(5 - 2x) = 2(3 - x^2)$$

$$5x - 2x^2 = 6 - 2x^2$$

$$5x = 6$$

The terms involving x^2 cancelled out, and $5x = 6$ *is* a linear equation with solution $x = \dfrac{6}{5}$. You can check to make sure that this is in fact the solution.

Example H   Solve $\dfrac{n+3}{n-2} = \dfrac{8}{3}$.

Multiply by $3(n - 2)$. You get

$$3(n + 3) = 8(n - 2)$$

$$3n + 9 = 8n - 16$$

$$5n = 25$$

$$n = 5$$

Check: $\dfrac{n+3}{n-2} = \dfrac{5+3}{5-2} = \dfrac{8}{3}$.

Example I   Solve $\dfrac{2}{\dfrac{4}{2r+3} - 3} = 2$.

Multiply the numerator and the denominator of the left side of the equation by $2r + 3$. This gives

$$\frac{2(2r + 3)}{4 - 3(2r + 3)} = 2$$

$$\frac{4r + 6}{4 - 6r - 9} = 2$$

$$\frac{4r + 6}{-6r - 5} = 2$$

$$4r + 6 = -12r - 10$$

$$16r = -16$$

$$r = -1$$

Be sure to check this solution in the original equation.

B5.4 PROBLEMS

Solve for the indicated variable.

1. $\dfrac{2}{x-3} + 1 = 0$ _____

2. $3 + \dfrac{1}{2-x} = 0$ _____

3. $5 - \dfrac{4}{3-2s} = 3$ _____

4. $\dfrac{6}{x-3} = 1$ _____

5. $\dfrac{1}{2x+3} + \dfrac{3}{2x-1} = 0$ _____

6. $\dfrac{1}{t} - \dfrac{1}{2} = \dfrac{2}{t}$ _____

7. $\dfrac{3}{n} = 1 + \dfrac{5}{2n}$ _____

8. $\dfrac{2}{3k} = -3 + \dfrac{8}{3k}$ _____

9. $\dfrac{1}{5+8w} = \dfrac{1}{4w-3}$ _____

10. $\dfrac{4}{5z+5} + \dfrac{2}{10z-1} = 0$ _____

11. $\dfrac{12x-1}{8x+1} + 2\left(\dfrac{3}{8x+1} + 1\right) = 0$ _____

12. $\dfrac{8}{3s} + \dfrac{3}{2s} = 1$ _____

13. $\dfrac{2}{3r} + \dfrac{3}{24r-5} = \dfrac{1}{3r}$ _____

14. $\dfrac{2}{x-4} - \dfrac{15}{x} = 0$ _____

15. $9(4 - x + x^2) = x(9x + 2)$ _____

16. $\dfrac{4}{5 + \dfrac{2}{3t}} = \dfrac{3}{4}$ _____

17. $\dfrac{4}{15} + \dfrac{1}{6k} - 1 = 0$ _____

18. $2x(3x - 4) = 6(8 + x^2)$ _____

19. $\dfrac{3}{1 + \dfrac{3}{2s}} = 2$ _____

20. $\dfrac{6b - 1}{1 - 6b} + 5b = 9$ _____

21. $\dfrac{2 - 4t}{2t - 1} - 6t = 10$ _____

22. $\dfrac{9}{1 + m} - 3\left(1 - \dfrac{2}{m + 1}\right) = 0$ _____

23. $\dfrac{21}{4x - 3} - \dfrac{4}{3 - 4x} = \dfrac{12}{8x - 1}$ _____

24. $\dfrac{5}{\dfrac{3}{2r - 1} + 2} = 6$ _____

25. $\dfrac{3}{5 - \dfrac{2}{z + 1}} = -4$ _____

26. A man drives n miles at 40 miles per hour, then drives the next n miles at 60 miles per hour. What is his average speed for the entire $2n$ miles? (It is not 50 miles per hour.) Remember that rate $= \dfrac{\text{distance}}{\text{time}}$.

B5.5 SOLUTION OF A LINEAR INEQUALITY IN ONE UNKNOWN WITH NUMERICAL COEFFICIENTS

In this unit, you will be dealing with **inequalities** rather than equalities. There are four inequality symbols:

$<$ means   "is less than"

$>$ means   "is greater than"

$\leq$ means   "is less than or equal to"

$\geq$ means   "is greater than or equal to"

Example A

a. $6 < 7$ b. $-2 < 19$

c. $7 > 6$ d. $19 > -2$

e. If $a \leq b$, then $b \geq a$

f. $(-5)^2 > 0$ g. $x^2 \geq 0$ for any real number x

There are some rules for inequalities that must be learned.

> RULE I
> You can add or subtract the same number on each side of an inequality, and the inequality will be maintained.

Example B

a. Since $4 < 7$, then $4 + 8 < 7 + 8$ because $12 < 15$.

b. If $x \leq 3$, then $x - 1 \leq 3 - 1$, so $x - 1 \leq 2$.

c. If $5x > 4$, then $5x - 2 > 2$.

d. If $3x - 4 \geq 6$, then $3x \geq 10$.

> RULE II
> You can multiply and divide each side of an inequality by the same *positive* number, and the inequality will be maintained.

Example C

a. Since $4 < 7$, then $4 \cdot 2 < 7 \cdot 2$ because $8 < 14$.

b. If $x \leq 3$, then $8x \leq 24$. (Both sides got multiplied by 8, which is positive.)

c. If $5x > 4$, then $x > \dfrac{4}{5}$. (Both sides got divided by 5, which is positive.)

d. If $3x - 4 \geq 6$, then $3x \geq 10$ (by Rule I) and so $x \geq \dfrac{10}{3}$ (by Rule II).

> RULE III
> You can multiply or divide each side of an inequality by the same *negative* number, but you must then reverse the inequality.

247

Example D

a. $-2 < 19$, so $(-2)(-1) > 19(-1)$ since $2 > -19$.

b. $6 < 7$, so $-6 > -7$.

c. If $5 - 2x > 11$, then $-2x > 6$ (by Rule I), and so $x < -3$ (by Rule III).

d. If $9 - 4x \leq 12 + x$, then $9 - 5x \leq 12$ (by Rule I), so $-5x \leq 3$ (by Rule I again), so $x \geq -\frac{3}{5}$ (by Rule III).

Two inequalities are called **equivalent** if you can get from one of them to the other by means of the above rules. To *solve* a linear inequality, you simply use the rules to find an equivalent inequality with the variable isolated on one side of the inequality sign.

Example E Solve the inequality $5t + 3 < 6 - 2t$.

$5t + 3 < 6 - 2t$	Add $2t$, Rule I
$7t + 3 < 6$	Subtract 3, Rule I
$7t < 3$	Divide by 7, Rule II
$t < \dfrac{3}{7}$	This is the solution

Example F Solve $4x - 7 \geq 6x - 11$.

$4x - 7 \geq 6x - 11$	Subtract $6x$, Rule I
$-2x - 7 \geq -11$	Add 7, Rule I
$-2x \geq -4$	Divide by -2, Rule III
$x \leq 2$	This is the solution

Example G Solve $4x - 7 \geq 6x - 11$ in another way.

$4x - 7 \geq 6x - 11$	Subtract $4x$, Rule I
$-7 \geq 2x - 11$	Add 11, Rule I
$4 \geq 2x$	Divide by 2, Rule II
$2 \geq x$	The same solution

Example H Solve $3(5 + 2x) < 16x - 9$, and in the solution, put the x on the left.

$$3(5 + 2x) < 16x - 9$$

$$15 + 6x < 16x - 9$$

$$24 + 6x < 16x$$

$$12 + 3x < 8x$$

$$12 < 5x$$

$$x > \frac{12}{5}$$

B5.5 PROBLEMS

In each blank, fill in one of the symbols $<$, $>$, or $=$ to make the statement true.

1. 5____8

2. $\dfrac{8}{4}$____$\dfrac{12}{3}$

3. $\dfrac{14}{2}$____$\dfrac{9}{3}$

4. 7____2

5. 2^3____$\dfrac{16}{2}$

6. $11 - 2$____3^2

7. $9 + 3$____$16 - 4$

8. $3\dfrac{1}{2}$____$\dfrac{7}{3}$

9. $4\dfrac{3}{8}$____$4\dfrac{1}{2}$

10. $\dfrac{111}{3}$____37

11. 4^2____$(-2)^4$

12. $(-3)^3$____$(-2)^5$

13. -12____-5^2

14. $6 - N$____$-N$

15. $a + 19$____$-8 + a$

16. -4^3____-18

17. $x(x - 2)$____$(x - 1)^2$

18. 2^{10}____10^3

19. 3^4____9^2

20. $(t + 1)(t - 1)$____$t^2 - 1$

Solve each inequality for the indicated variable.

21. $3 + 2x < 9$_____

22. $4 - 6x > -8$_____

23. $7 - 5x > -18$_____

24. $5 + 3x < 17$_____

25. $3y - 4 \le 5y$_____

26. $3(n - 2) > 0$_____

27. $5(r + 3) < 0$_____

28. $5t - 27 \le 2t$_____

29. $3S - 4 \le S$_____

30. $15 - 4x \ge x$_____

249

31. $32 - 3t \geq 5t$ _____

32. $7T + 5 \leq 2T$ _____

33. $3(2 - 3r) < 4 + r$ _____

34. $2(2x - 5) \geq 3(2 - x)$ _____

35. $5(4 - y) \geq 3(4y - 2)$ _____

36. $4(5 - 3s) < s + 6$ _____

37. $\frac{2}{3}x - 4 \leq \frac{1}{2}x + 1$ _____

38. $t(2 + 2t) > 2t(t - 2) - 3$ _____

39. $3n(2 + n) > n(1 + 3n) + 2$ _____

40. $\frac{3}{4}y + 2 \leq \frac{1}{3}y - 5$ _____

41. $\frac{7}{3}k - \frac{3}{4} < \frac{2}{5} - \frac{5}{2}k$ _____

42. $\frac{3}{2}\left(\frac{3}{2}x - 3\right) > \frac{x}{5}$ _____

43. $\frac{r}{7} > \frac{5}{3}\left(\frac{2}{7}r - 4\right)$ _____

44. $\frac{3}{8} + \frac{9}{5}r < \frac{4}{3}r - \frac{2}{3}$ _____

B5.6 SOLUTION OF TWO LINEAR EQUATIONS IN TWO UNKNOWNS WITH NUMERICAL COEFFICIENTS — BY SUBSTITUTION

You have seen how to solve a single linear equation in one unknown. In this unit, you will solve two linear equations in two unknowns.

Example A If both $3x - y = 4$ and $5x + 2y = 3$ are true, what must x and y be?

This is called a system of equations, and we write it as

$$3x - y = 4$$

$$5x + 2y = 3$$

From the first equation, solve for y. Thus, $y = 3x - 4$. Now substitute this value for y in the second equation. This yields

$$5x + 2(3x - 4) = 3$$

$$5x + 6x - 8 = 3$$

$$11x = 11$$

$$x = 1$$

Now, since $x = 1$ and $y = 3x - 4$, it must be true that

$$y = 3 \cdot 1 - 4$$

$$= -1$$

The solution to this system of equations is $x = 1$, $y = -1$. Let us check that these values for x and y satisfy both equations.

$$3(1) - (-1) = 3 + 1 = 4$$

$$5(1) + 2(-1) = 5 - 2 = 3$$

Example B Solve the system

$$2x - 4y = 4$$

$$x - y = 5$$

From the second equation, $x = 5 + y$. Then from the first equation,

$$2(5 + y) - 4y = 4$$

$$10 + 2y - 4y = 4$$

$$10 - 2y = 4$$

$$6 = 2y$$

$$y = 3$$

251

Now substitute $y = 3$ in the second equation (because that's easier).

$$x - 3 = 5$$
$$x = 8$$

The solution to this system of equations is $x = 8$, $y = 3$. Let us check that these values for x and y satisfy both equations.

$$2(8) - 4(3) = 16 - 12 = 4$$
$$8 - 3 = 5$$

Example C   Solve the system

$$2x + 3y - 5 = 0$$
$$3x - 2y = 6$$

From the first equation, you get

$$x = \frac{5 - 3y}{2}$$

Now substitute this into the second equation:

$$3\left(\frac{5 - 3y}{2}\right) - 2y = 6$$

$$\frac{15 - 9y}{2} - 2y = 6$$

$$15 - 9y - 4y = 12$$

$$15 - 13y = 12$$

$$13y = 3$$

$$y = \frac{3}{13}$$

Next, use this value of y in either equation to find x. Let's use the second equation.

$$3x - 2\left(\frac{3}{13}\right) = 6$$

$$3x - \frac{6}{13} = 6$$

$$39x - 6 = 78$$

$$13x - 2 = 26$$

$$13x = 28$$

$$x = \frac{28}{13}$$

The solution to the system is $x = \dfrac{28}{13}$, $y = \dfrac{3}{13}$. You should check that these values satisfy both of the original equations.

Example D   The sum of John's age and Mary's age is 48. John is 12 years older than Mary. How old are they?

Let

$\quad J =$ John's age

$\quad M =$ Mary's age

The first sentence can be expressed as

$\quad J + M = 48$

The second sentence can be expressed as

$\quad J = M + 12$

Thus, we need to solve this system of equations:

$\quad J + M = 48$

$\qquad J = M + 12$

Now substitute J from the second equation into the first equation.

$$J + M = 48$$
$$(M + 12) + M = 48$$
$$2M + 12 = 48$$
$$2M = 36$$
$$M = 18$$

Then from the second equation,

$$J = M + 12$$
$$= 18 + 12$$
$$= 30$$

John is 30 years old, and Mary is 18.

You can check that these ages fit the data of the problem.

Example E   One number is 4 more than another. Their sum is 3 times their difference. What are the numbers?

Let

x = the larger number

y = the other number

Then the first sentence can be written as

$x = y + 4$

The second sentence can be written

$x + y = 3(x - y)$

Thus, the two equations to be solved are

$x = y + 4$

$x + y = 3(x - y)$

Substituting x from the first equation into the second equation yields

$y + 4 + y = 3(y + 4 - y)$

$2y + 4 = 3 \cdot 4$

$2y + 4 = 12$

$2y = 8$

$y = 4$

Since $y = 4$, then

$x = y + 4$

$= 4 + 4$

$= 8$

The solution is $x = 8$, $y = 4$. Again, you should check the solution for the conditions of the problem.

B5.6 PROBLEMS

Use substitution to solve each system of equations for the indicated variables.

1. $x = 4y + 1$

    $4x + 3y = 4$

2. $a = b + 1$

    $2a - 3b = 1$

3. $r = s - 1$

    $3r - s = 3$

4. $y = 3x + 2$

    $4x + 2y = 4$

5. $n = 2m + 7$

    $m + n = 10$

6. $v = 1 - u$

    $u + 2v = 5$

7. $d = 3 - 2c$

    $4c + d = -1$

8. $q = 5p - 15$

    $2p + q = -1$

9. $2x + y = 5$

    $4x - 3y = -7$

10. $4s + t = 6$

    $8s - 3t = -3$

11. $2p - q = -2$

    $6p - 2q = -1$

12. $x - 3y = 1$

    $2x + 9y = 7$

13. $2x - 3y = 7$

    $3x + 2y = 5$

14. $6u + 3v = -4$

    $5u - 2v = 8$

15. $3a - 5b = 11$

$4a - 3b = -3$

16. $4x - 5y = 9$

$2x + 3y = 4$

17. $2m - 7n + 8 = 0$

$5m + 2n = 12$

18. $4r - 5s = -2$

$3r + 2s + 9 = 0$

Solve each problem by writing two equations in two unknowns, then solving them by substitution.

19. Polly and Esther together have \$50. Tomorrow, Polly will receive \$5 and Esther will spend \$7. Then Polly will have twice as much money as Esther. How much does each have now?

20. Mary Kay is 5 years older than Jay. The sum of their ages is 57. How old is each?

21. At a rock concert, the cost for tickets was \$5 for students and \$8 for others. There were 772 paid admissions, and the gate receipts totaled \$5996. How many students were at the concert?

22. There were 3691 ballots cast in an election. If 13 votes had switched from the winner to the loser, the loser would have won by one vote. How many votes did each get?

B5.7 SOLUTION OF TWO LINEAR EQUATIONS IN TWO UNKNOWNS WITH NUMERICAL COEFFICIENTS — BY ELIMINATION

In this unit, another method will be used to solve a system of two linear equations. It is done by eliminating one of the variables. The examples will reveal the method.

Example A Solve the system

$$2x + 3y = -5$$

$$4x - 3y = 17$$

One very handy way to solve this system is to add the corresponding terms of each equation:

$$\begin{array}{r} 2x + 3y = -5 \\ 4x - 3y = 17 \\ \hline 6x = 12 \\ x = 2 \end{array}$$

Now $x = 2$ can be substituted into either equation. Let's use the first one.

$$2 \cdot 2 + 3y = -5$$

$$3y = -9$$

$$y = -3$$

Let us check that the pair $x = 2$, $y = -3$ does indeed satisfy both equations.

$$2(2) + 3(-3) = 4 - 9 = -5$$

$$4(2) - 3(-3) = 8 + 9 = 17$$

Example B Solve the system

$$5A + 4B = 1$$

$$3A + 2B = -1$$

This time, just adding or subtracting won't get rid of either variable. But if the terms of the second equation are all multiplied by 2, then the equations become

$$5A + 4B = 1$$

$$6A + 4B = -2$$

Now B can be eliminated by subtracting the first equation from the second. The result is $A = -3$. Now this value can be substituted back into either equation to get $B = 4$. You should check in the original equations that this pair solves the system.

Example C   Solve the system

$$2r - 3t = 1$$
$$3r + 2t = 21$$

This time, multiply the terms of the first equation by 2. Also, multiply the terms of the second equation by 3. The result is

$$4r - 6t = 2$$
$$9r + 6t = 63$$

Now add corresponding terms of the two equations. The result is

$$13r = 65$$
$$r = 5$$

Since $r = 5$, the second original equation gives

$$3 \cdot 5 + 2t = 21$$
$$15 + 2t = 21$$
$$2t = 6 \quad t = 3$$

The solution is $r = 5$, $t = 3$. Check it out.

Example D   Solve the system

$$4x + 3y = 5$$
$$12x - 15y = -1$$

One way is to multiply the first equation through by 5. The system becomes

$$20x + 15y = 25$$
$$12x - 15y = -1$$

Now add, to get

$$32x = 24$$
$$x = \frac{24}{32} = \frac{3}{4}$$

Use this in the first equation.

$$4 \cdot \frac{3}{4} + 3y = 5$$
$$3 + 3y = 5$$
$$3y = 2 \quad y = \frac{2}{3}$$

Check to make sure that $x = \frac{3}{4}$ and $y = \frac{2}{3}$ solves both equations.

B5.7 PROBLEMS

Use elimination to solve each system.

1. $x + y = 5$

   $x - y = 1$

2. $2x + 6y = 18$

   $3x + 6y = 21$

3. $5r - 3s = 4$

   $r - 3s = 6$

4. $a - b = 9$

   $a + b = -1$

5. $m + 2n = 6$

   $3m + 5n = 18$

6. $4J + 3K = 0$

   $9J - 2K = -35$

7. $5u - 4v = 0$

   $9u - 7v = -2$

8. $3p + q = -13$

   $2p - 3q = 6$

9. $6A + 5B = 61$

   $5A - 6B = 0$

10. $2x - 3y = 33$

    $5x + 4y = 48$

11. $3s + 5t = 14$

    $6s - 3t = 67$

12. $4P + 3Q = 7$

    $3P - 4Q = -26$

13. $8x - 7y = 12$

    $5x + 3y = 4$

14. $6a - 5b = 8$

    $9a + 3b = -4$

15. $2m + 9n = 14$

    $5m - 4n = 12$

16. $8s + 3t = 5$

    $5s - 7t = -6$

Use either method you wish—substitution or elimination—to solve each system.

17. $a - b = 5$

    $a + 3b = 8$

18. $2m - 9n = 4$

    $3m + 3n = 5$

19. $4R - 3S = 7$

    $2R + 5S = 1$

20. $x - 4y = 7$

    $x + 2y = 2$

21. $3x - 5y = 8$

    $2x + 3y = 5$

22. $9t - 5u = -2$

    $4t + 2u = 5$

23. $7m + 3n = 8$

    $5m - 4n = 6$

24. $4p - 3q = 8$

    $3p - 5q = -2$

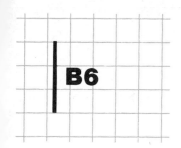

B6

SOLUTION OF QUADRATIC EQUATIONS

6.1 SOLUTION OF A QUADRATIC EQUATION FROM FACTORED FORM

Quadratic equations are equations that contain a squared term and no terms to a higher power, such as $x^2 = 36$, $x^2 - 4 = 0$, and $x^2 + 3x - 10 = 0$. These are also known as second-degree equations. Quadratic equations may have no solution, one solution, or two solutions.

Example A Find x when $x^2 = 25$.

The value of x could be either $+5$ or -5, since both equal 25 when squared.

There is still another way of looking at this quadratic equation. Here we set the quadratic equation equal to 0 and solve it by factoring. For example, $x^2 = 25$ can be written as $x^2 - 25 = 0$. This can be factored into $(x - 5)(x + 5) = 0$. You now see that if either $(x - 5)$ or $(x + 5)$ is 0, you have a solution to the quadratic. Thus, for

$x - 5 = 0$ x must be equal to 5

and for

$x + 5 = 0$ x must be equal to -5

Whenever a quadratic equation is set equal to 0 and then factored, it may be possible to find the solutions by setting each factor equal to 0 and solving it, just as you would solve a linear equation. This is all based on the fact that if P and Q are numbers and $P \cdot Q = 0$, then either $P = 0$ or $Q = 0$.

Example B Given $(x + 2)(x - 3) = 0$.

What is the quadratic equation and the solutions to it?

267

To get the quadratic equation, multiply the factors.

$$(x + 2)(x - 3) = x^2 - x - 6$$

To find the solutions, set each factor equal to 0, and solve each as a linear equation.

$$x + 2 = 0 \qquad x = -2$$
$$x - 3 = 0 \qquad x = 3$$

The two solutions are: (-2) and (3). Check each by substituting in the original equation.

$$-2: \qquad (x + 2)(x - 3) = 0$$
$$(-2 + 2)(-2 - 3) = 0$$
$$0 = 0 \qquad -2 \text{ checks}$$

$$3: \qquad (x + 2)(x - 3) = 0$$
$$(3 + 2)(3 - 3) = 0$$
$$0 = 0 \qquad 3 \text{ checks}$$

When given the factors for a quadratic equation, we can easily find the solutions by setting each factor equal to 0 and solving for the variable.

Example C Find the solutions for $(2a - 8)(a + 7) = 0$.

First factor: $\quad 2a - 8 = 0 \quad 2a = 8 \quad a = 4$
Second factor: $\quad a + 7 = 0 \qquad\qquad\qquad a = -7$
Check: $\qquad\quad (2a - 8)(a + 7) = 0$
Substitute 4: $\quad (8 - 8)(4 + 7) = 0$
$$0 = 0 \qquad\qquad\qquad 4 \text{ checks}$$
Substitute -7: $\quad (-14 - 8)(-7 + 7) = 0$
$$0 = 0 \qquad\qquad\qquad -7 \text{ checks}$$

Example D Find the solutions to $(3x + 12)(2x + 10) = 0$.

Set each factor equal to 0 and solve.

$$3x + 12 = 0 \quad 3x = -12 \quad x = -4$$
$$2x + 10 = 0 \quad 2x = -10 \quad x = -5$$

To check, substitute values in the original equation.

In summary, when there are two solutions to a quadratic equation, they can be found by first factoring the quadratic and then setting each factor equal to 0 and solving. The check should always be performed to insure you have not made a mistake somewhere.

B6.1 PROBLEMS

Find and check solutions to each of the following factored quadratics.

1. $(x - 2)(x + 3) = 0$

Solutions _____

Quadratic _____

Check _____

2. $(2a - 4)(a + 5) = 0$

Solutions _____

Quadratic _____

Check _____

3. $(a - 6)(2a + 6) = 0$

Solutions _____

Quadratic _____

Check _____

4. $(2x + 3)(4x - 5) = 0$

Solutions _____

Quadratic _____

Check _____

5. $(7b - 14)(7b + 14) = 0$

Solutions _____

Quadratic _____

Check _____

6. $(5q + 10)(5q - 10) = 0$

Solutions_____

Quadratic_____

Check_____

7. $(x - 6)(x + 6) = 0$

Solutions_____

Quadratic_____

Check_____

8. $(9p + 3)(9p - 3) = 0$

Solutions_____

Quadratic_____

Check_____

9. $(2m - 2)(2m + 2) = 0$

Solutions_____

Quadratic_____

Check_____

10. $(4w + 1)(w - 1) = 0$

Solutions_____

Quadratic_____

Check_____

11. $(6 + w)(8 + w) = 0$

Solutions_____

Quadratic_____

Check_____

12. $(x + 2)(2 + x) = 0$

Solutions_____

Quadratic_____

Check_____

B6.2 SOLUTION OF A QUADRATIC EQUATION BY FACTORING

We have seen that when presented with a quadratic equation that has been factored, it is quite easy to find its solutions. Simply set each factor equal to 0 and solve. However, when given a quadratic equation to solve that has not been factored, we have a more formidable problem.

The first step is to get the equation into a form that may be factorable. This will involve getting all terms on one side of the equality and setting the equation equal to 0. It may also be possible to factor out common terms.

Example A Solve $a^2 - 4a = 5$.

To factor this quadratic, first rewrite, or subtract 5, from both sides. Thus,

$$a^2 - 4a = 5 \quad \text{becomes} \quad a^2 - 4a - 5 = 0$$

This can be factored.

$$a^2 - 4a - 5 = 0$$
$$(a - 5)(a + 1) = 0$$

Set each factor equal to 0 and solve.

$$a - 5 = 0 \quad a = 5$$
$$a + 1 = 0 \quad a = -1$$

There are two solutions: $a = 5$ and $a = -1$.

The two solutions can be checked by substituting each into the original equation to see if they work.

Using $a = 5$: $\quad (5)^2 - 4(5) = 5$

$\qquad\qquad\qquad 25 - 20 = 5 \quad \text{yes}$

Using $a = -1$: $\quad (-1)^2 - 4(-1) = 5$

$\qquad\qquad\qquad\quad 1 + 4 = 5 \quad \text{yes}$

The key to solving (if it is possible) a quadratic by factoring it is to begin by setting the quadratic equal to 0. Divide out any factor that is common to *all* terms; factor; set each factor equal to 0 and solve.

Finally, check your solutions by replacing them in the original equation.

Example B Find solutions for $3x^2 + 15x = -18$.

$$3x^2 + 15x + 18 = 0 \qquad \text{Set it equal to 0}$$

$$3(x^2 + 5x + 6) = 0 \qquad \text{Factor out 3}$$

$$x^2 + 5x + 6 = 0 \qquad \text{Divide by 3}$$

$$(x + 2)(x + 3) = 0 \qquad \text{Factor the quadratic}$$

Set each factor equal to 0 and solve.

$$x + 2 = 0 \qquad x = -2$$

and

$$x + 3 = 0 \qquad x = -3$$

Check each of the two solutions (-2) and (-3) by substituting them into the original equation.

Example C Find solutions for $2a^2 = 18$.

$$2a^2 - 18 = 0$$

Factor the greatest common factor, 2.

$$2(a^2 - 9) = 0$$

Divide by 2.

$$a^2 - 9 = 0$$

Factor.

$$(a - 3)(a + 3) = 0$$

Solve.

$$a - 3 = 0 \qquad a = 3$$

$$a + 3 = 0 \qquad a = -3$$

Check each solution in the original equation.

B6.2 PROBLEMS

Solve each of the following by first factoring. Check each solution.

1. $x^2 + 2x + 1 = 0$

Factors _____

Solutions _____

Checks _____

2. $x^2 - 25 = 0$

Factors _____

Solutions _____

Checks _____

3. $a^2 - 4a = -4$

Factors _____

Solutions _____

Checks _____

4. $b^2 - b - 6 = 0$

Factors _____

Solutions _____

Checks _____

5. $2x^2 - 4x = 6$

Factors _____

Solutions _____

Checks _____

6. $25x^2 + 5x = 2$

Factors _____

Solutions _____

Checks _____

7. $2w^2 + 4w = 16$

Factors _____

Solutions _____

Checks _____

8. $6p^2 + 13p = -6$

Factors _____

Solutions _____

Checks _____

9. $-x^2 + 6x - 9 = 0$

Factors _____

Solutions _____

Checks _____

10. $18r^2 + 6r = 4$

Factors _____

Solutions _____

Checks _____

11. $-25x^2 + 20x = 4$

Factors _____

Solutions _____

Checks _____

12. $6a^2 + 12a = -6$

Factors _____

Solutions _____

Checks _____

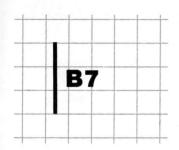

GRAPHING

B7.1 GRAPHING A POINT ON THE NUMBER LINE

The use of a number line greatly assists us in understanding the various kinds of numbers and to locate points on lines. Observe the following:

Whole number line:

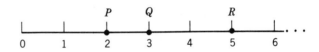

Example A Point P is located at number 2.

Point Q is located at number 3.

Point R is located at number 5.

Integer number line:

Example B Point X is located at number -3.

Point Y is located at number $+2$.

Point Z is located at number 0.

Rational number line:

275

Example C   Point A is located at number $-\dfrac{1}{8}$.   Point C is located at number $\dfrac{5}{8}$.

Point B is located at number $\dfrac{1}{8}$.   Point D is located at number 1.

Referring to the fraction (rational) number line above, you see that there are many possible fractions that could be inserted between any two given fractions. Indeed, there are an infinite number.

Example D   On the rational number line of Example C, between which two numbers would you find $\dfrac{5}{7}$?

There are several ways this could be solved.

> One way: Since the line is marked in eighths, find a common multiple of 7 and 8. We can use 56. Convert $\dfrac{5}{7}$ to $\dfrac{40}{56}$. Now you can see that $\dfrac{5}{7}$ is between $\dfrac{5}{8}$, or $\dfrac{35}{56}$, and $\dfrac{3}{4}$, or $\dfrac{42}{56}$.
>
> Another way: Convert the fractions to decimals. This is done by dividing numerator by denominator. $\dfrac{5}{7} = 0.714285\ldots$. Since this fraction does not terminate, it is a matter of how far we need to carry the division to locate the point. $\dfrac{5}{8} = 0.625$ and $\dfrac{3}{4} = 0.75$. Again, we see $\dfrac{5}{7}$ as falling between $\dfrac{5}{8}$ and $\dfrac{3}{4}$.

Example E   On the number line below, which letter best locates the number 0.269?

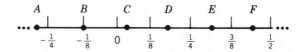

First, estimate about where you think it might be; perhaps $\dfrac{3}{8}$?

$\dfrac{3}{8}$ as a decimal = 0.375   Too big

Try $\dfrac{1}{4}$. $\dfrac{1}{4} = 0.25$. Too small. But we now know our number lies between $\dfrac{1}{4}$ and $\dfrac{3}{8}$. The letter best locating this is E.

Example F   Using the same number line, which letter best locates -0.158?

Let us estimate $-\dfrac{1}{8}$. As a decimal, it equals -0.125. If we estimate $-\dfrac{1}{4}$, we get -0.25. Which letter is closer to -0.158? B.

Locating points on a number line can help in understanding relationships between numbers.

B7.1 PROBLEMS

On the following number line, which letter best locates each fraction?

1. $\dfrac{1}{2}$ _____

2. $\dfrac{3}{4}$ _____

3. $\dfrac{4}{7}$ _____

4. $\dfrac{5}{6}$ _____

5. $\dfrac{4}{5}$ _____

6. $\dfrac{5}{4}$ _____

7. $\dfrac{1}{16}$ _____

8. $\dfrac{6}{5}$ _____

9. $\dfrac{1}{10}$ _____

10. $\dfrac{5}{12}$ _____

Using the same number line, which letter best locates each decimal?

11. 0.25 _____

12. 0.59 _____

13. 0.45 _____

14. 0.789 _____

15. 0.125 _____

16. 1.05 _____

On the number line following, which letter best locates each number?

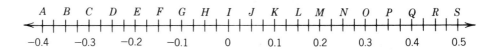

17. 0.25 _____

18. 0.33 _____

19. 0.375 _____

20. −0.15 _____

21. 0.29 _____

22. $-\dfrac{2}{7}$ _____

277

23. 0.35 _____   24. 0.48 _____

25. −0.35 _____   26. −0.01 _____

27. −0.30 _____   28. 0.005 _____

29. −0.20 _____   30. $\frac{5}{12}$ _____

31. −0.42 _____   32. 0.001 _____

B7.2 GRAPHING LINEAR INEQUALITIES
IN ONE UNKNOWN

Using number lines to illustrate various numerical relationships is extremely valuable. The essential concepts are: "equal to," "less than," and "greater than." On the number line, their meaning is apparent. Study the number line and the examples relating to it. Several numbers are already graphed and are represented by the indicated points and letters.

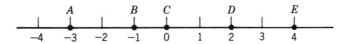

Example A If we refer to a number x that is equal to 4, we are talking about the point located at letter E. If we say x is equal to -1, we are talking about the point located at letter B.

Example B If we refer to the numbers "x less than 2," we are talking about all numbers to the left of 2, but not including the number 2. On our preceding number line, it is all the numbers to the left of letter D. As a graph on the number line, it looks like this:

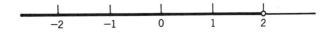

Notice that the point representing number 2 is open; this indicates that it is not included in numbers less than 2.

Example C If we refer to a number "x greater than -3," we are talking about all numbers to the right of -3, but not including -3.

Again notice that the point representing -3 is open, meaning that it is not included in numbers greater than -3.

We can now begin to combine our graphing skills by using the words "OR" and "AND". In general, the word "or" means that a number is included if it belongs to one set of conditions or the other. It is best illustrated on the number line.

Example D Graph the numbers that are "greater than 2" OR "less than 0."

We first look at the number line and graph the numbers greater than 2. We then graph the numbers less than 0. The final graph is illustrated below.

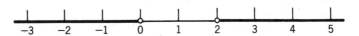

279

Any number x that is greater than 2 or less than 0 fits the conditions. This includes numbers like -18, -5, $-\dfrac{5}{7}$, 3, 29, $30\dfrac{1}{2}$, 50, etc.

The other relation we can best show by a graph is "AND". It differs from "or" in that all conditions must be met to be included in the graph.

Example E   Graph the numbers that are "less than 5" AND "greater than -2."

They are shown in the graph below.

Note that neither 5 nor -2 is included, but all the numbers in between, such as $-\dfrac{3}{2}$, -1, $-\dfrac{2}{3}$, 0, $\dfrac{3}{4}$, 1, etc.

When graphing, care must be taken to keep clearly in mind the difference between the words "or" and "and". They are frequently used in graphing.

B7.2 PROBLEMS

Make a graph of each of the following on the given number lines. Remember that the symbol $<$ means "less than" and that $>$ means "greater than."

1. $x > 2$

2. $x < 5$

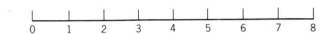

3. $x < -1$

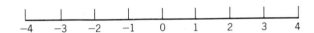

4. $x < \dfrac{3}{4}$

5. $x > 0.3$

6. $x < 9$ and $x > 1$

7. $x > -2$ and $x < 4$

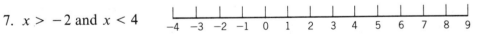

8. $x < -1$ or $x > 3$

9. $x < \dfrac{1}{4}$ or $x > \dfrac{5}{8}$

10. $x > -0.1$ and $x < 0.6$

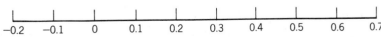

281

Study each of the following graphs and then write conditions to describe it.

11. _____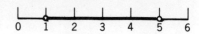

12. _____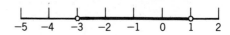

13. _____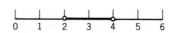

14. _____

15. _____

16. _____

17. _____

18. _____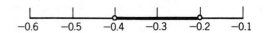

19. _____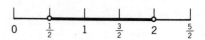

20. _____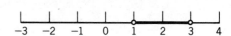

B7.3 GRAPHING A POINT IN THE COORDINATE PLANE

We have seen how to graph, and locate, points on number lines. We can also use numbers to locate points in a plane. But since a plane, illustrated by a flat surface such as a sheet of paper, has two dimensions, we need to use two numbers. Essentially we use two number lines, perpendicular to each other, and crossing at the number 0 on each line. Observe the diagram below.

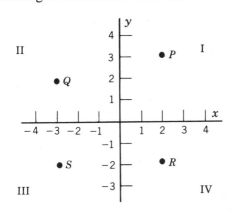

As you see, we have two number lines, the horizontal one is labeled, and named, the "x" direction. The vertical dimension is labeled "y".

Also note that there are some points in the diagram that are labeled P, Q, R, and S. We will describe where they are located by the use of two numbers. This pair of numbers is called the "coordinates of the point." It is important to learn that the first number given will always be the one for the horizontal or x direction, the second is for the vertical, or y direction. The coordinates for a point are also called "ordered pairs," because the order makes a difference.

Example A   What are the coordinates for point P?

Looking at the diagram, we see point P is located 2 units in the x direction and 3 units in the y direction. Point P is located at coordinates $(2, 3)$.

Example B   What are the coordinates of point Q?

We find letter Q and see it is located in the x direction at -3 and in the y direction at 2. Thus, the coordinates for point Q are $(-3, 2)$.

Example C   What are the coordinates of point R?

In the x direction, we have 2. In the y direction, we have -2. Thus, point R has coordinates $(2, -2)$.

Example D   What are the coordinates of point S?

Point S is located at -3 in the x direction and -2 in the y direction. Point S has coordinates $(-3, -2)$.

283

You have perhaps observed that by dividing the plane with the two number lines, we have made four separate sectors. These are called **quadrants**, and in quadrant I, the coordinates are all positive numbers. In quadrant II, the x coordinates are negative and the y coordinates are positive. In quadrant III, both coordinates are negative. And in quadrant IV, x is positive and y is negative.

Example E In the diagram to the right, what can you say about the coordinates of point A?

$x > 0$ and $y > 0$

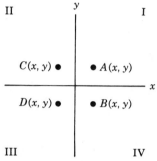

Example F What can you say about the coordinates of point C?

$x < 0$ and $y > 0$

Example G What about coordinates of points B and D?

Point B: $x > 0$ and $y < 0$

Point D: $x < 0$ and $y < 0$

Graphing in the coordinate plane (sometimes called the Cartesian Coordinate System) is a powerful tool in mathematics. It ties together the subjects of geometry, arithmetic, and algebra, and helps us visualize many concepts.

B7.3 PROBLEMS

Using the graph, give the approximate coordinates for each point to the nearest half unit.

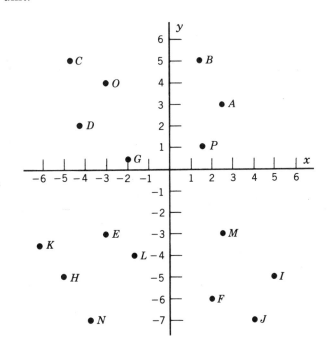

1. Point A _____   2. Point B _____

3. Point C _____   4. Point D _____

5. Point E _____   6. Point F _____

7. Point G _____   8. Point H _____

9. Point I _____   10. Point J _____

11. Point K _____   12. Point L _____

13. Point M _____   14. Point N _____

15. Point O _____   16. Point P _____

Using the coordinate system indicated, plot each point according to its given coordinates.

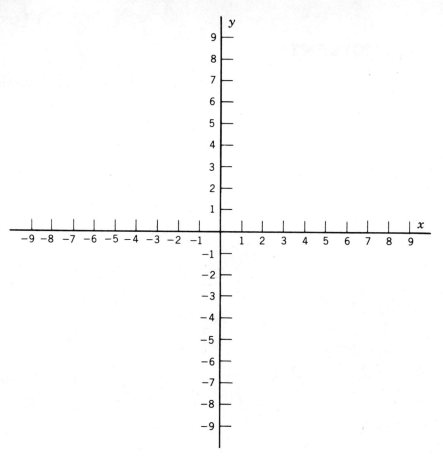

17. $A(3, 4)$

18. $B(-3, 5)$

19. $C(-7, -4)$

20. $D(4, -1)$

21. $F(4, -1)$

22. $G(6, -4)$

23. $H(-3, -7)$

24. $I(-9, 3)$

25. $J(3, -9)$

26. $K(8, -8)$

27. $L(-1, 0)$

28. $M(8, -2)$

29. $N(-7, 3)$

30. $O(0, 0)$

31. $P(-5, -5)$

32. $Q(0, -8)$

33. $R(5, -5)$

34. $S(2, -9)$

35. $T(-4, 8)$

36. $U(0, 8)$

B7.4 GRAPHING A SIMPLE LINEAR EQUATION:
$y = mx,\ y = b,\ x = b$

The technique for plotting points in a plane using ordered pairs of numbers (coordinates) can be expanded to include plotting lines, which are determined by *linear equations*.

Consider the points with the following coordinates:

They all lie on this line:

Point A $(0, 0)$

Point B $(1, 1)$

Point C $(2, 2)$

Point D $(-1, -1)$

Point E $(-2, -2)$

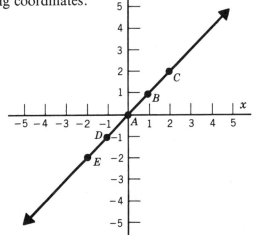

Notice that the x-coordinate is exactly the same as the y-coordinate for each point.

The line that passes through all these points is called the **graph of a linear equation**, and for this set of points it is written as the equation $y = x$.

Example A   Draw the graph of the equation $y = 2x$.

The first thing to do is to find some points to plot. Using a table like the one shown is helpful. Pick any value for the x-coordinate and solve the equation for the y-coordinate.

x	y
1	2
2	4
0	0
-1	-2
$\dfrac{1}{2}$	1

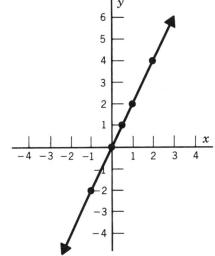

The line through these points is the graph of the linear equation $y = 2x$.

If you compare it to the first equation, $y = x$, you will notice $y = 2x$ is "steeper." What is your guess about the "steepness" of $y = 5x$? This steepness or tilt of a line is called the **slope** of the line.

In Example A, we plotted several points, to illustrate what happens, but we could have found the line by plotting just two points and drawing a line through them. This line would be the graph of all the ordered pairs defined by that linear equation.

Example B   Graph $y = -2x$, using just two points.

Make a table as before. Pick any two values for x, and solve for y.

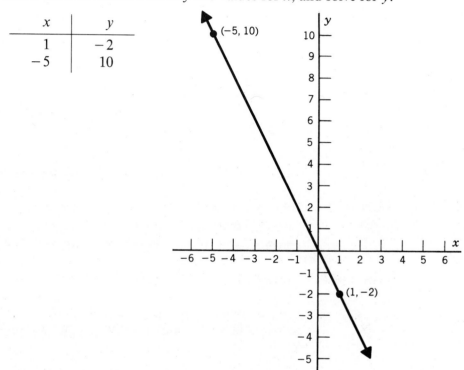

x	y
1	-2
-5	10

Compare this equation $y = -2x$ with the equation $y = 2x$ from Example A. One slopes to the right and the other to the left.

Example C   Graph $y = \frac{1}{4}x$, using 2 points.

Select two values for x and solve for y.

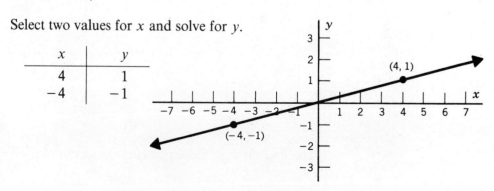

x	y
4	1
-4	-1

In writing linear equations, such as the preceding, notice that sometimes a number is written before the x (when it is not written, it is understood to be 1). This number, the **coefficient** of x, is the value that denotes the slope of the line. The larger this number is, the "steeper" the line will be. If the coefficient has a negative sign, it means the line will slope upwards to the left instead of to the right, as shown in Example B. The general form for a linear equation through $(0, 0)$ is given by $y = mx$, where m (a number) is the slope.

Example D   Graph $2y = 3x$.

First solve for y: $y = \dfrac{3}{2}x$. Select two values for x and solve for y.

x	y
2	3
-2	-3

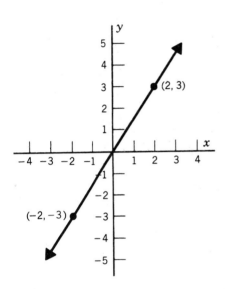

Example E   Graph $y = 2$.

This linear equation indicates that for all values of x, $y = 2$. It says nothing about x, which means x can be any number. So the graph is a line parallel to the x-axis, passing through 2 on the y-axis.

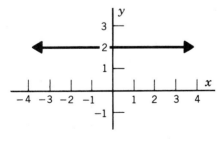

Example F   Graph $x = -2$.

This will be a line parallel to the y-axis passing through -2 on the x-axis.

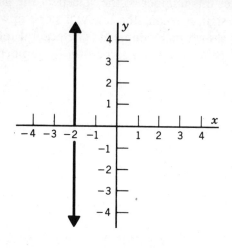

B7.4 PROBLEMS

For each of the following linear equations, find four points and plot each line on the grid.

1. $y = 5x$

x	y

2. $y = -3x$

x	y

3. $y = -\dfrac{4}{5}x$

x	y

4. $y = \dfrac{3}{2}x$

x	y

Write each of the following in the form for graphing. Determine two points and graph.

5. $3x = y$

x	y

291

6. $2y = -8x$

x	y

7. $\frac{1}{4}y = \frac{3}{4}x$

x	y

8. $5x = -2y$

x	y

9. $x = 3$

x	y

10. $y = -4$

x	y

11. $3 = 2y$

x	y

12. $7 = -3x$

x	y

B7.5 READING DATA FROM A GRAPH

Much of the information we use comes from data that has been put into graphic form. This makes it possible for us to see the information quickly and to appreciate how it may be related to other data.

The format of graphs varies considerably. Some even get quite cute, with various kinds of pictures in different sizes. But all have the same intent: to convey information rapidly.

Example A Study the following **bar graph**.

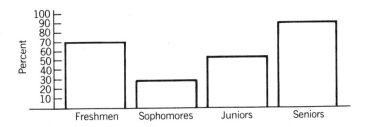

The graph represents the percentage of students, by school year, who are working part-time while attending their university. Note that the vertical direction represents percent and the horizontal is the university grade level.

a. In what grade level is the smallest percentage working?

By looking at the graph, you can readily see it is the sophomore class, with only 30%. This is the kind of data you can get directly by reading the graph.

b. If the university has 3500 seniors, how many seniors work part time?

Here we need to make a calculation based on the information given in the graph. Since 90% of seniors work, compute 3500 × .90, which equals 3150 students.

c. If there are 5000 freshmen, and 70% work, how many freshmen do not work?

Here, again, we need to perform a calculation based on data in the graph. Since 70% work, 30% do not work. Multiply 5000 by .30, and you see that 1500 freshman students do not work.

Bar graphs of the kinds shown in Example A are frequently used because of their simplicity and ease of reading.

Another graph frequently encountered is the **circle graph**. Here the data given is shown as a proportional part of the total circle. It may be used to show the relative percent of each category. Since the circle represents the entire collection of data, it represents 100%.

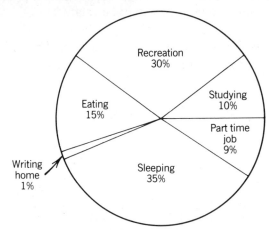

Example B   This circle graph represents the percentage of time a student at the university spends at various daily tasks.

a. What percent of time is devoted to:

    1. Studying? 10%
    2. Eating? 15%
    3. Sleeping? 35%

b. Since there are 24 hours in a day, how many hours are spent:

    1. Studying? $24 \times 0.1 = 2.4$ hours
    2. Recreation? $24 \times 0.3 = 7.2$ hours
    3. Writing home? $24 \times 0.01 = 0.24$ hours ≈ 14 minutes

Graphs of the kinds shown here convey information quickly and understandably, but sometimes, we must make additional computations in order to answer questions and make decisions.

Example C   If the university student is indeed using his days as illustrated in Example B and flunking several courses, where should he perhaps make some time adjustments?

By looking at the circle graph, the student may be embarrassed by the fact that he spends 30% of time on recreation and only 10% on studying. (A shift in percentages might improve the grades, eh?)

As you read magazines and newspapers, try to find various graphs and see what they are really telling you and also what they are not.

B7.5 PROBLEMS

Problems 1–6 refer to the bar graph used in Example A.

Assume that there are 5000 freshmen, 4250 sophomores, 4000 juniors, and 3500 seniors enrolled in the university.

1. How many sophomores are not working? _____

2. How many juniors are working? _____

3. How many freshmen are working? _____

4. What percent of the students who are working are upper classmen (juniors or seniors)? _____

5. What percent of the students who are working are lower classmen (freshmen or sophomores)? _____

6. What percent of the total student population works? _____

Problems 7–12 refer to the circle graph used in Example B.

7. What percent of time does the student devote to recreation or sleeping?

_____

8. What percent of time does the student spend *not* sleeping? _____

9. How many hours during a 24-hour day does the student devote to studying or eating? _____

10. How many hours per day does the student devote to studying or eating or sleeping? _____

11. If the student decreases the number of hours per day spent on recreation to 5 and used the remaining time to study, what percent of time does the student now devote to studying? _____

12. If the student quits the job and divides that time equally between studying and recreation, what percent of time does the student now devote to recreation? _____

Problems 13–18 refer to this graph of grades made in a mathematics course of several sections.

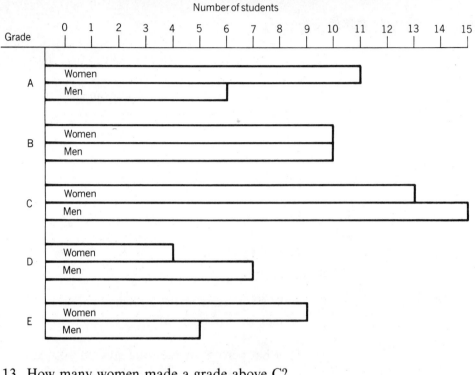

13. How many women made a grade above C?_____

14. How many men made a grade below B?_____

15. How many women were in the class?_____

16. How many students made an unsatisfactory grade (D or F)?_____

17. How many women made a grade of C or better?_____

18. What percent of the students made below C?_____

19. What percent of the men made C or better?_____

20. How many men were in the class?_____

GEOMETRIC MEASUREMENT SKILLS

C1

MEASUREMENT FORMULAS FOR PERIMETER AND AREA OF TRIANGLES, SQUARES, RECTANGLES, AND PARALLELOGRAMS

C1.1 MEASUREMENT: LINE SEGMENTS AND ANGLES

Measurement is a major concern of humankind. We measure our weight, our height, the temperature, the amount of sugar in a recipe, popularity of a political candidate, ratings of a television program, depth of a swimming pool, amount of pollution, direction of the wind, our blood pressure, and so on, in every aspect of our lives.

The measurements mentioned above are approximations. In fact, almost all measurements are approximations (try to think of one that is not).

In this unit, you will measure line segments and angles. To measure a line segment, use a ruler that is scaled in inches and in centimeters. To measure the length of line segment AB, place the ruler so that A is lined up with a mark on the scale of the ruler. Then observe the mark that B is lined up with. The number of inches (or centimeters) and parts of inches (or centimeters) between the marks associated with A and with B is called the **length** of segment AB.

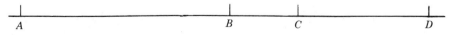

Figure C1-1

Make sure you see that

1. $AB = 2.25$ in.
2. $BC = 2$ cm
3. $AD = 11.2$ cm
4. $BD = 5.5$ cm
5. $AC = 3$ in.
6. $CD = 3.7$ cm

305

In each measurement, there is room (and reason for) error. It is surely not true that *AC* is *exactly* 3 inches, yet we write *AC* = 3 in. In some books, you'll find other notations to mean that *AC* is *approximately* 3 in. We will use the equal sign, with the understanding that if a measurement is involved, it is automatically an approximation.

In measuring lengths, you see that the more divisions of units there are, the more accurate the measurements. Rulers scaled in centimeters are usually divided into tenths of a centimeter. Now 1 in. = 2.54 cm, so a sixteenth of an inch is 0.15875 cm, which is greater than a tenth of a centimeter. Thus, measurements can be made more accurately on the centimeter scale than on the inch scale.

To measure an angle, you will use a protractor, shown in Figure C1-2. A protractor is scaled in **degrees**, defined as follows: If $\angle BAC$ (the notation for "angle *BAC*") has the property that *A*, *B*, and *C* all lie on the same line, with *A* between *B* and *C*, then $\angle BAC$ has a measure of 180 degrees, each degree being an equal part of these 180 degrees.

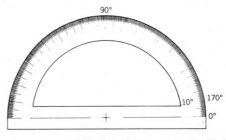

Figure C1-2

To measure an angle, it is necessary to denote just what is to be measured. This is commonly done with an arrow, as shown in Figure C1-3. In this figure, the arc designates that the number of degrees in $\angle XYZ$ is 37. We write $m(\angle XYZ) = 37°$ (the superscript "°" meaning "degrees") and this is read, "The measure of angle *XYZ* is 37 degrees." This is measured by the protractor as shown in Figure C1-4. This angle may be designated $\angle XYZ$, $\angle ZYX$, or simply $\angle Y$.

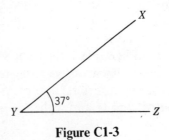

Figure C1-3

The center point at the bottom of the protractor is placed at the vertex of the angle. Then one of the sides of the angle is lined up to show 0°, and the other gives the reading of 37°. As with segments, the measurement of angles entails error, and we use the equality sign in the same approximate manner. That does not keep us from naming some angles in terms of the exactness of degrees:

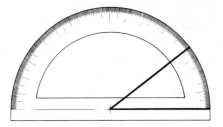

Figure C1-4

An **acute** angle is an angle of more then 0° but less than 90°;

A **right** angle is an angle of exactly 90°;

An **obtuse** angle is an angle of more than 90° but less than 180°;

A **straight** angle is an angle of exactly 180°.

Example A   Figure C1-5 shows some angles.

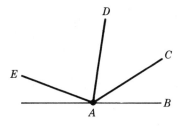

Figure C1-5

Make sure you see that

1. $m(\angle BAC) = 32°$ 4. $m(\angle CAD) = 49°$

2. $m(\angle BAD) = 81°$ 5. $m(\angle CAE) = 128°$

3. $m(\angle BAE) = 160°$ 6. $m(\angle DAE) = 79°$

Degrees are very small units. But in astronomy, engineering, and many other fields, more accurate measurements are needed. One way to do this is to use decimal parts of degrees, as in 73.2° or 19.41°.

Angle measures are expressed in degrees. Degrees are divided into minutes and seconds as follows:

1 degree = 60 minutes = 60′

1 minute = 60 seconds = 60″

Fractional parts of degrees can also be expressed in decimal form. This is done by dividing minutes by 60.

Example B $15' = \dfrac{15}{60}$ degrees $= 0.25°$

$50' = \dfrac{50}{60}$ degrees $= 0.8\overline{3}°$

Example C Use the fact that 1 in. $= 2.54$ cm, exactly: (a) Convert $3\dfrac{3}{4}$ in. to centimeters; (b) convert 8.6 cm to inches.

a. $3\dfrac{3}{4}$ in. $= \left(3\dfrac{3}{4} \times 2.54\right)$ cm

$= (3.75 \times 2.54)$ cm

$= 9.5250$ cm

$= 9.53$ cm (rounded)

b. 8.6 cm $= (8.6 \div 2.54)$ in.

$= 3.385826771\ldots$ in.

$= 3.39$ in. (rounded)

C1.1 PROBLEMS

1. Measure the right-hand edge of this page to the nearest

   a. inch _____

   b. half-inch _____

   c. quarter-inch _____

   d. centimeter _____

   e. millimeter (tenth of a centimeter) _____

2. a. Draw a segment that you think is 4 in. long. Now measure the length of that segment. How much was your error? _____

   b. Repeat part (a) for 7 in. _____

   c. Repeat part (a) for 9 in. _____

   d. Repeat part (a) for 6 cm. _____

   e. Repeat part (a) for 8 cm. _____

   f. Repeat part (a) for 13 cm. _____

3. a. Measure each of these five segments in inches, as accurately as you can, to the nearest unit your ruler shows. Record your results just above each segment.

   _____

   _____

   _____

   _____

   _____

   b. Now measure those five segments in centimeters and millimeters (tenths of centimeters). Record your results above each segment.

c. Convert each measurement in part (a) to centimeters, using the fact that 1 in. = 2.54 cm, exactly. Record your results above each segment.

d. Compare your results in part (c) with those in part (b). What do you notice?_____

4. Use your protractor to draw an angle of

   a. 17°   b. 80°

   c. 135°   d. 143°

5. a. Draw an angle that you think is 45°. Now measure that angle with your protractor. How much was your error?_____

   b. Repeat part (a) for 60°._____

   c. Repeat part (a) for 30°._____

   d. Repeat part (a) for 85°._____

   e. Repeat part (a) for 100°._____

   f. Repeat part (a) for 115°._____

6. Convert to centimeters (two decimal places).

   a. 3.7 in._____   b. 18.43 in._____

   c. 3 ft 11$\frac{1}{2}$ in._____   d. 15$\frac{5}{8}$ in._____

   e. 2 yd 2 ft 2 in. *248.92*   f. 7$\frac{11}{16}$ in._____

7. Convert to inches (two decimal places).

   a. 1016 cm_____   b. 41.78 cm_____

   c. 17 cm_____   d. 43.6 cm_____

   e. 143.3 cm_____   f. 1.257 m_____

8. Convert to decimal degrees (two places).

   a. 43° 15′_____   b. 187° 3′_____

   c. 176° 42′_____   d. 217° 13′_____

C1.2 PERIMETER

The perimeter of a figure is the sum of the lengths of all of its sides. For some figures, it may be possible to find shortcuts to compute the perimeter. This is true when more than one side has the same length.

Example A   Find the perimeter P of this triangle:
$P = 5 + 8 + 12 = 25$.

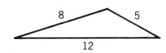

In general, the perimeter for a triangle is $P = a + b + c$, where a, b, c, are the lengths of the sides.

Example B   Find the perimeter of this square:
$P = 3 + 3 + 3 + 3 = 12$.

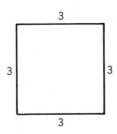

The general formula for the perimeter of a square is $P = 4s$, where s is the length of a side.

Example C   Find the perimeter of this rectangle:
$P = 8 + 8 + 4 + 4 = 24$.

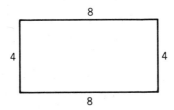

311

The general formula for the perimeter of a rectangle is $P = 2\ell + 2w$, where ℓ is the length and w is the width.

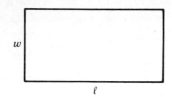

Example D Find the perimeter of this regular hexagon:
$P = 3 + 3 + 3 + 3 + 3 + 3 = 18$.

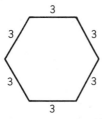

The general formula for the perimeter of a regular hexagon is $P = 6s$, where s is the measure of a side.

Note: We have used the word **regular**. This means that all sides and all angles have the same measure.

In Examples B, C, and D, we were able to use a formula to assist in finding the perimeters, because they were regular polygons.

Example E Find the perimeter of this parallelogram (parallelograms have two pairs of parallel and equal sides; squares and rectangles are special parallelograms):
$P = 2(4) + 2(6) = 20$.

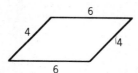

Although two parallelograms might have the same perimeter, they could look quite different.

Example F Find the perimeter of this polygon (note that it is not regular):
$P = 3 + 5 + 6 + 7 + 2 = 23$

The general formula is $P = a + b + c + d + e$.

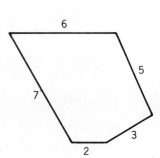

In Example F, there is no shortcut formula:
You just add the lengths of the sides.

C1.2 PROBLEMS

Find the perimeter for each of the following.

8.4

1.
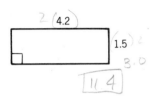
2 (4.2)
1.5
3.0
11 4

2.
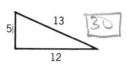
13   30
5
12

3.
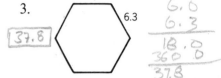
37.8   6.3
6.0
6.3
18.0
360 0
37.8

4.
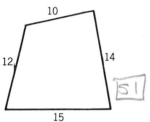
10
12   14
51
15

5.
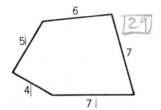
6   29
5   7
4
7

6.
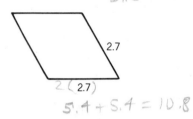
2.7
2 (2.7)
5.4 + 5.4 = 10.8

7.
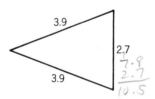
3.9
3.9
2.7
7.8
2.7
10.5

8.
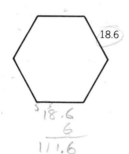
18.6
18.6
6
111.6

9.
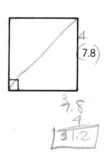
7.8
7.8
4
31.2

10.
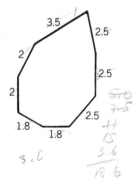
3.5
2.5
2
2.5
2
2.5
1.8
1.8
3.6
7.8
+
15
3.6
18.6

The following polygons are named. State the formula for finding the perimeter and then compute that perimeter for each polygon.

11. Rectangle

7   14
3   6

12. Square

5
Sx4=20

Formula   Perimeter   Formula   Perimeter

2l+2W   20   P=4S   20   **313**

20

13. Triangle

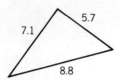

7.1   5.7   8.8

Formula   Perimeter

P = A + B + C   21.6

14. Triangle

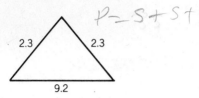

P = S + S + S

2.3   2.3   9.2

Formula   Perimeter

P = A + B + C   13.8

15. Regular octagon (8 sides)

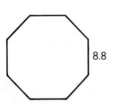

8.8

Formula   Perimeter

P = 8S   70.4

16. Regular hexagon

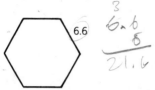

6.6

3
6.6
8
21.6

Formula   Perimeter

P = 6S   21.6

17. Parallelogram

10.2
5.10
4.6   5.2

2l + 2w

Formula   Perimeter

2l + 2w   15.4

18. Rectangle

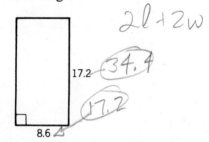

2l + 2w

17.2   34.4

17.2

8.6

Formula   Perimeter

P = 2l + 2w   51.6

19. Square

4s

23.2

Formula   Perimeter

P = 4S   92.8

20. Parallelogram

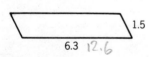

1.5
6.3   12.6

Formula   Perimeter

P = 2l + 2w   15.6

1
8.8
7.1
5.7
21.6

2.3
2.3
4.6
9.2
13.8

6
8.8
8
70.4

5.1
2
10.2

4.6
2
5.2

17.2
2
34.4
17.2
51.6

23.2
4
92.8

C1.3 AREA

Every polygon has an interior region. It is the shaded part of each of the following polygons.

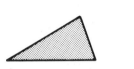 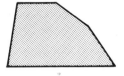

The area of a polygon is the number of area units contained in its interior. An area unit is a square representing the unit of measurement used. It could be in square inches, square centimeters, square yards, square kilometers, etc.

Example A Given this square (it could be in any unit desired) find the area of each polygon.

Area unit = ☐

 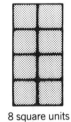 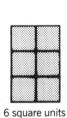 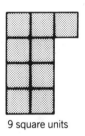

4 square units   8 square units   6 square units   9 square units

There are many formulas that help in finding areas of various polygons. Here are some, developed in a sequence that shows how the formulas are related.

Example B Find the area A of rectangle $PQRS$ by using the formula $A = \ell \times w$, where ℓ = length and w = width.

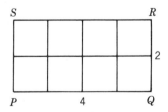

The length is 4 and the width is 2. Area = $4 \times 2 = 8$ square units. Count the squares contained in the rectangle. Again, you get 8 square units.

315

Example C   Find the area of parallelogram $TUVW$. Use the formula $A = b \times h$, where $b = $ base and $h = $ height.

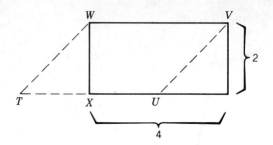

The base is a side on which the altitude or height is drawn.
Notice that this was done by cutting off triangle TXW and moving it to the other side, thus forming a rectangle. The height of the parallelogram becomes the width of the rectangle.
$A = (4)(2) = 8$.

This formula will also work for the rhombus, since it is a special parallelogram where all sides are of equal length.

Example D   Find the area of triangle ABC by using the formula $A = \frac{1}{2}\ell \times w$.
$A = \frac{1}{2}(4)(2) = 4$.

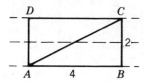

Notice that the triangle ABC was formed by dividing the rectangle into two equal parts with a diagonal.

Example E   Find the area of triangle ABC using the formula $A = \frac{1}{2}\ell \times w$.

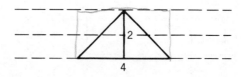

Notice that the triangle looks different from Example D. However, it has the same area.

Example F   Find the area A of trapezoid $HIJK$ using the formula $A = \frac{1}{2}h(B + b)$ where h = height, B = one base, and b = the other base. A trapezoid is four-sided figure with exactly two parallel sides.
$A = \frac{1}{2}(2)(5 + 3) = \frac{1}{2}(2)(8) = 8.$

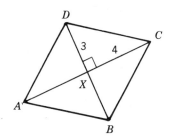

Example G   A rectangle has perimeter 18 and width 4. What is its area?

In rectangles, we know that $P = 2\ell + 2w$ and $A = \ell \times w$. This means that $18 = 2\ell + 2(4)$ or $18 = 2\ell + 8$. Thus, $\ell = 5$. We can now compute the area: $A = \ell \times w = 5 \times 4 = 20.$

Example H   In the rhombus shown, the two diagonals meet at right angles, $AC = 8$, $CX = 4$, and $DX = 3$. Find the area.

Note that there are four triangles, all with equal area. We use the area formula for triangles, $A = \frac{1}{2}bh$.

$A = \frac{1}{2} \cdot 4 \cdot 3 = 6$

So the total area is $4 \cdot 6 = 24.$

C1.3 PROBLEMS

Find the area for each of the following.

1. Area =

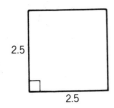

2. Area =

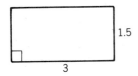

3. Area =

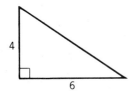

4. Area =

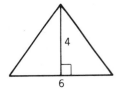

5. Area =

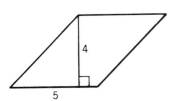

6. Area =

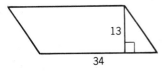

7. Area =

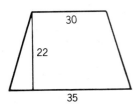

8. Area =

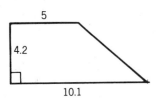

9. Area =

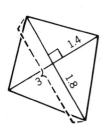

10. Area =

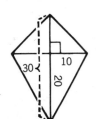

Solve each of the following:

11. A rectangle has perimeter 24 and width 4. What is the area?

12. A triangle has base 12 and height 4. What is the area?

13. A rectangle has area 30 and width 5. What is the length?

14. The area of a triangle is 15. If the base is 5, what is the height?

15. Right triangle ABC has area 20 with the right angle at A. If $AB = 8$, what is AC?

16. A rectangle has area 32. If the length is twice the width, what are the length and the width?

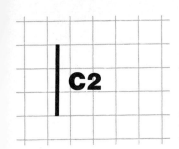

MEASUREMENT FORMULAS FOR CIRCUMFERENCE AND AREA OF CIRCLES

C2.1 MEASUREMENT FORMULAS FOR CIRCUMFERENCE OF CIRCLES

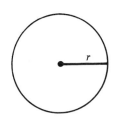

The perimeter of a circle is called its **circumference**. The circumference is the product of 2 times π (pi) and the radius r of the circle. (The **radius** of a circle is the distance from the circle's center to any point on the circle.) The formula for finding circumference is:

$$C = 2\pi r$$

The Greek letter π (pi) is used to denote the ratio of the circumference of a circle to its diameter. Since the **diameter** d of a circle is twice its radius, an equivalent formula for finding the circumference is:

$$C = \pi d$$

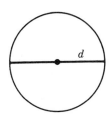

The decimal value of the symbol π (pi) cannot be written as a finite or even an infinite repeating decimal. We will use 3.14 as an approximation for that value. Greater accuracy is possible by using more decimal places. (pi = 3.141592...) Pi is the number that represents the ratio of a circle's circumference C to its diameter d. Diameter $d = 2r$.

$$\pi = \frac{C}{d}$$

Example A Find the circumference of a circle with radius 5.

$$C = 2\pi r \approx 2(3.14)(5) = 31.4$$

or

$$C = \pi d \approx (3.14)(10) = 31.4$$

where $\approx$ means approximate equality.

321

Finding the circumference of a circle is like finding the perimeter of a polygon. However, the number π is always involved, and this introduces a certain inaccuracy when computation with decimals takes place.

Example B   If the circumference of a circle is 10 in., what is its radius? (To the nearest hundredth of an inch.)

$$C = 2\pi r$$

$$r = \frac{C}{2\pi}$$

$$r = \frac{10}{2(3.14)}$$

$$r \approx 1.59 \text{ in.}$$

The ratio of the circumferences of two circles is equal to the ratio of the lengths of their radii (radii is plural for radius). In the two circles following, this proportion is expressed as:

$$\frac{C}{C'} = \frac{r}{r'}$$

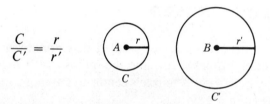

Example C   The radius of circle A is 18 and the radius of circle B is 24. What is the ratio of their circumferences?

$$\frac{C}{C'} = \frac{18}{24} = \frac{3}{4}$$

The ratio of C to C' is 3 to 4.

Example D   In Example C, what is C' if $C = 27.6$?

$$\frac{27.6}{C'} = \frac{3}{4} \quad \text{or} \quad C' = \frac{4(27.6)}{3} = 36.8$$

Example E   When the circumference of circle A is twice the circumference of circle B and the radius of circle A is 6, what is the radius of circle B?

Let the radius of circle B be r'. Then

$$\frac{C}{C'} = \frac{r}{r'}$$

$$\frac{2}{1} = \frac{6}{r'}$$

$$r' = 3$$

C2.1 PROBLEMS

Find the circumference and/or perimeter for each figure. Use 3.14 for π.

1. Circle of radius 8 $C =$ _____

2. Circle of radius 5.2 $C =$ _____

3. Perimeter of square $ABCD$ $P =$ _____

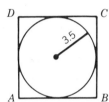

4. Circumference of the small
   circle shown $C =$ _____

5. Circumference of the
   circle shown $C =$ _____

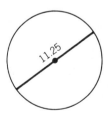

6. Perimeter of the shape
   shown $P =$ _____

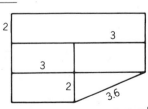

323

7. Circumference of inscribed circle.
   Area of square is 144. $C =$ _____

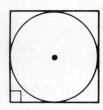

8. Perimeter of the shape
   shown $P =$ _____

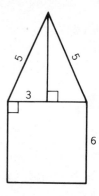

Find each answer. Round off when appropriate.

9. If the circumference of a circle is 18 in., what is its radius? _____

10. If the circumference of circle A is four times the circumference of circle B,
    and the radius of circle B is 3, what is the radius of circle A? _____

11. If the circumference of circle A is one-half the circumference of circle B,
    and the radius of circle A is 6, what is the radius of circle B? _____

12. Circle A has radius 8, and circle B has radius 24. What is the ratio of their
    circumferences? _____

13. Find the perimeter of the shape at
    the right. _____

14. Circle A has circumference 28.
    What is its diameter? _____

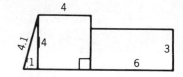

C2.2 MEASUREMENT FORMULAS FOR AREA
OF CIRCLES

The area of a circle is the product of the square of its radius and the number π. We will again use 3.14 for π, remembering that this is only an approximation.

$$A = \pi r^2$$

Example A   In circle y, the radius is 6. What is its area?

$$A = \pi r^2 \qquad A = \pi (6)^2 \approx (3.14)(36) = 113.04$$

Example B   A circle has circumference 6π. What is its area? In this situation, we must use two formulas:

$$A = \pi r^2 \quad \text{and} \quad C = 2\pi r$$

First, we solve for r in $C = 2\pi r$. Since $6\pi = 2\pi r$, then $r = 3$.

Now substitute 3 for r in $A = \pi r^2$.

$$A = \pi (3)^2 \approx 3.14 \times 9 = 28.26$$

The ratio of the areas of two circles is equal to the ratio of the squares of their radii. This relationship may be stated as follows.

$$\frac{\text{Area of circle } A}{\text{Area of circle } B} = \frac{r^2}{r'^2}$$

Example C   If the radius of circle A is 2 and the radius of circle B is 3, what is the ratio of their areas?

$$\frac{\text{Area of circle } A}{\text{Area of circle } B} = \frac{2^2}{3^2} = \frac{4}{9}$$

Sometimes we may need to use several formulas to solve a problem.

Example D   Find the distance around and area of the figure below with the information given (a square with two semicircles).

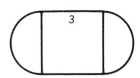

The circumference of the two semicircles is equal to the circumference of a complete circle.

1. To find the circumference of the circle:

$$C = 2\pi r \approx 2(3.14)\left(\frac{3}{2}\right) = 9.42$$

The length of the remaining two sides is $2(3) = 6$.

Total distance $\approx 9.42 + 6 = 15.42$.

2. To find the area:

Area of the two semicircles is equal to the area of a complete circle.

$$A = \pi r^2 \approx (3.14)\left(\frac{3}{2}\right)^2 = 3.14 \times \frac{9}{4} = 7.065 \approx 7.07.$$

Area of square $=$ (side)(side)

$$A = s^2 = 3^2 = 9$$

Total area $\approx 7.07 + 9 = 16.07$.

Finding circumferences and areas of circles has involved the number π. It introduces a degree of error whenever we use a decimal value for it. However, we have no other choice.

Example E   The area of a circle is 10 square units. What is its radius?

$$A = \pi r^2$$

$$10 = 3.14 r^2$$

$$r^2 = \frac{10}{3.14} \approx 3.18$$

$$r = \sqrt{3.18} \approx 1.8$$

The radius is about 1.8 units. (A calculator is helpful here.)

C2.2 PROBLEMS

Complete the following table for each circle. Use 3.14 for π.

	Radius	Diameter	Circumference	Area
1. Circle A	8	_____	_____	_____
2. Circle B	_____	6	_____	_____
3. Circle C	_____	38	_____	_____
4. Circle D	_____	_____	42	_____
5. Circle E	_____	_____	_____	50
6. Circle F	1	_____	_____	_____
7. Circle G	_____	4	_____	_____
8. Circle H	_____	_____	600	_____
9. Circle I	_____	_____	_____	1
10. Circle J	_____	14	_____	_____

Give each of the following.

11. Find the area of a circle whose circumference is 24π.

12. The circumference of circle A is four times the circumference of circle B. What is the radius of circle B? The diameter of circle A is 16.

13. If the enclosed circle in the figure has a radius of 6.5, what is the area of the square?

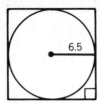

14. What is the perimeter of the figure? Each end is a semicircle.

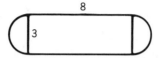

15. If the radius of circle A is 4 and the radius of circle B is 8, what is the ratio of their areas?

16. Find the radius of a circle whose circumference is numerically equal to its area.

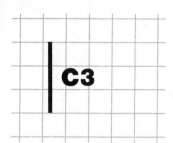

C3
MEASUREMENT FORMULAS FOR VOLUME

C3.1 MEASUREMENT FORMULAS FOR VOLUME OF CUBES AND RECTANGULAR SOLIDS

To find volume (V), we measure how much a container holds (capacity), or how much space an object takes up. As with distance and area, we begin by selecting a unit of volume. A cube with one unit of length on each edge has a volume of one cubic unit.

The size units we may wish to use depend upon our tasks; they could be inches, centimeters, kilometers, miles, etc.

Example A Illustrate 1 cubic inch (1 in.^{3}).

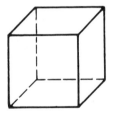

To find the volume of a rectangular solid, we must determine the number of cubic units it contains.

Example B Find the volume of the rectangular solid shown.

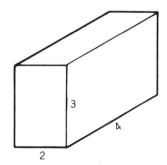

329

Imagine filling this container with 1-cm cubes. How many will it take? For this, we can use the formula

$$V = \ell \times w \times h \quad \ell = \text{length}, \ w = \text{width}, \ h = \text{height}$$

$$V = 4 \times 2 \times 3 = 24 \ \text{cm}^3 = 24 \ \text{cubic centimeters}$$

Example C   Find the volume of the following rectangular solid:

$$V = \ell \times w \times h \qquad V = (2.5)(1.5)(3.5) = 13.125$$

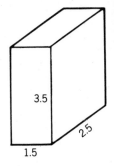

Notice that the formula works even when the measures of the sides are not whole numbers.

Cubes are special cases of the rectangular solids because all the sides have the same length. Thus, the formula for finding volume of a cube is simply

$$V = s^3$$

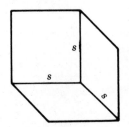

Example D   If the side of a cube is 4 in., what is the volume?

$$V = (4)^3 = 64 \ \text{cubic inches}$$

Example E   By how much do you increase the volume of a cube if you double the length of each side?

Cube A: Volume $= 1^3$ Cube B: Volume $= 2^3 = 8$

The volume is increased eightfold!

Example F   A cube has a volume of 27 units. What are its dimensions?

$$V = 27$$

We now seek a number that when cubed equals 27.

$$V = s^3 \qquad 27 = s^3 \qquad s = 3$$

Thus, a cube with 27 cubic units measures 3 by 3 by 3.

Finding the volume of rectangular solids is a matter of applying the formula $V = \text{length} \times \text{width} \times \text{height}$.

C3.1 PROBLEMS

Find the volume for each.

1. Rectangular solid:

   $\ell = 8$, $w = 6$, $h = 5$

2. Rectangular solid:

   $\ell = 2.5$, $w = 3.5$, $h = 0.1$

3. Cube: $s = 4$

4. Cube: $s = 0.01$

5. Rectangular solid:

   $\ell = \dfrac{1}{2}$, $w = \dfrac{3}{4}$, $h = \dfrac{1}{8}$

6. Rectangular solid:

   $\ell = 2$, $w = \dfrac{1}{2}$, $h = 1$

7. Cube: $s = \dfrac{1}{5}$

Solve each problem.

8. If the length of a side of a cube is 6, what is the volume? _____

9. If the length of the sides of Problem 9 are doubled, what is the volume? ____

10. Given the rectangular solid shown,

    find its volume. _____

A   1.5   B

11. If in Problem 11, AB is tripled, what is the volume?_____

12. If you triple the length of a side in a rectangular solid, what will happen to the volume? Illustrate.

13. If you double the length of just one side in a rectangular solid, will the volume be doubled? Make two examples to demonstrate your answer.

14. Find the volume of the shape shown.

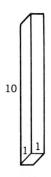

15. Find the volume of the shape shown.

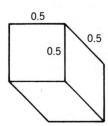

C3.2 MEASUREMENT FORMULAS FOR VOLUME OF CYLINDERS AND SPHERES

Finding the volume of shapes such as cylinders and spheres is similar to finding the volume of rectangular solids except it involves the number π.

In the case of cylinders, we find the area of the base, which is circular and therefore involves π, and then multiply that area times the height. The result is the volume expressed in whatever volume units are used: cubic inches, cubic centimeters, cubic feet, etc.

Example A   Find the volume of the cylinder shown.

1. Find the area of the base, which is a circle with radius 5.

$$A = \pi r^2 \approx 3.14(5)^2 = 78.5 \text{ square units}$$

2. Multiply the area of the base times the height 8.

$$V \approx 78.5 \times 8 = 628 \text{ cubic units}$$

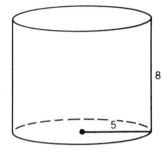

We now generalize a formula for finding volumes of cylinders:

$$V = \underset{\text{area of base}}{\pi r^2} \times \underset{\text{height}}{h}$$

$$= \pi r^2 h$$

Example B   Find the volume of a cylinder with a base radius of 3 ft and height of 7 ft.

$$V = \pi r^2 h \approx (3.14)(3)^2(7) = 197.82 \text{ ft}^3$$

Example C   A cylindrical water tank with radius 2 ft and height (length) 8 ft is half full of water. How many cubic feet of water does it contain?

Total $V = \pi r^2 h = \pi(2)^2 8 = 32\pi \text{ ft}^3$

One-half tank $= \dfrac{1}{2} \cdot 32\pi = 16\pi \text{ ft}^3$

Notice that in Example C we used π and did not convert it to a decimal. This is often done during computation, since it simplifies the work and also does not introduce any error.

333

The volume of a sphere is given by the formula

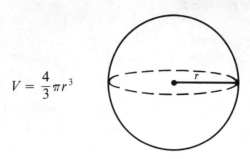

$$V = \frac{4}{3}\pi r^3$$

Example D Find the volume of a sphere with radius 2.

$$V = \frac{4}{3}\pi r^3 = \frac{4}{3}\pi (2)^3 = \frac{4}{3}\pi (8) = \frac{32}{3}\pi = 10\frac{2}{3}\pi \text{ cubic units.}$$

Example E A spherical water tank with a diameter of 20 ft contains how many cubic feet of water?

$$V = \frac{4}{3}\pi r^3$$

Since we are given a diameter of 20 ft, the radius is 10.

$$V = \frac{4}{3}\pi (10)^3 = 1333\frac{1}{3}\pi \text{ ft}^3$$

Example F If the water tank in Example E is $\frac{1}{3}$ full, how much water does it contain?

Full volume was $3333\frac{1}{3}\pi$ ft^3

$$\frac{1}{3}\left(\frac{4000\pi}{3}\right) = \frac{4000\pi}{9} = 444\frac{4}{9}\pi \text{ ft}^3$$

Finding the volumes of cylinders and spheres is a matter of keeping in mind that the number π will be involved. This presents no problem in computation and, in general, you will probably find leaving π in its symbolic form preferable to using the decimal form.

C3.2 PROBLEMS

Find the volume for each: Use π, not 3.14.

1. Cylinder with radius 5 ft, height 9 ft. $V =$ _____

2. Cylinder with radius 8.6 m, height 7.3 m. $V =$ _____

3. Sphere with radius 9 in. $V =$ _____

4. Sphere with radius 4.5 cm. $V =$ _____

5. Cylinder with radius 3 in., height 3 in. $V =$ _____

6. Sphere with radius 3 in. $V =$ _____

Solve.

7. How do the volumes of the cylinder in Problem 5 and the sphere in Problem 6 compare?

8. A water tank has a radius of 5 ft and a length of 9 ft. If it is $\frac{3}{4}$ full, how many cubic feet of water does it contain?

9. If you double the length of a cylinder, but leave the radius the same, what happens to the volume?

10. A sphere has a radius of 5. If you double this radius to 10, how much greater will the volume become?

11. If you triple the diameter of a sphere, how much will this increase the volume?

12. A cylinder has a base area of 10 in.^{2} If the volume is 80 in.^{3}, what is the height of the cylinder?

ACHIEVEMENT TEST

C1   MEASUREMENT FORMULAS FOR PERIMETER AND AREA OF TRIANGLES, SQUARES RECTANGLES AND PARALLELOGRAMS

C2   MEASUREMENT FORMULAS FOR CIRCUMFERENCE AND AREA OF CIRCLES

C3   MEASUREMENT FORMULAS FOR VOLUME

---

1. A rectangle has perimeter 16 and width 2. What is the area of the rectangle?

   (A) 4   (B) 12   (C) 16   (D) 20   (E) 32

---

2. A rectangle has an area of 32. If the width is 4, what is the perimeter?

   (A) 16   (B) 24   (C) 12   (D) 8   (E) 20

---

3. A rectangle with area of 64 and length of 16 has what width?

   (A) 4   (B) 6   (C) 8   (D) 10   (E) 12

4. The area of parallelogram *PQRS* shown is 36. *PQ* = 9. What is *TS*?

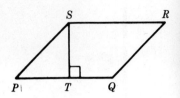

   (A) 4   (B) 6   (C) 9   (D) 12   (E) 15

5. In the figure shown, $AC \perp BD$, $AC = 6$, $EB = 3$, $ED = 5$. What is the area of quadralateral *ABCD*?

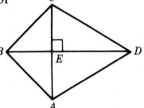

   (A) 30   (B) 16   (C) 20   (D) 32   (E) 24

6. The area of right triangle *XYZ* is 45. If $YZ = 9$, then $XY =$

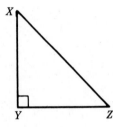

   (A) 3   (B) 5   (C) 15   (D) 10   (E) $\dfrac{45}{2}$

7. Rectangle *ABCD* has a length of 6 and width of 2. If you double both the length and the width, what will the area become?

   (A) 24   (B) 36   (C) 48   (D) 60   (E) 72

8. The area of triangle *ABC* shown is 60. If $AB = 20$, what is *XC*?

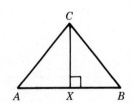

   (A) 6   (B) 3   (C) 30   (D) 10   (E) 15

9. The area of square *PQRS* shown is 16. What is the radius of the circle?

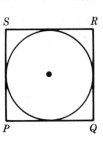

(A) 4   (B) 8   (C) 2π (D) 4π (E) 2

10. What is the area of a circle whose circumference is 16π?

(A) 8π (B) 64π (C) 16π (D) 256   (E) 32π

11. If the radius of a circle with radius 6 is doubled, what happens to the area?

(A) Doubles   (B) Increases by π (C) Quadruples

(D) Increases by 2π (E) Increases by $\dfrac{\pi}{2}$

12. A garden is in the shape shown to the right. *ABCD* is a square with side length of 6. The two ends are semicircles with radius 3. What is the perimeter of the garden?

(A) $12 + 3\pi$ (B) 6π (C) $12 + 2\pi$ (D) 24   (E) $6\pi + 12$

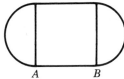

13. The cube shown has an edge of 6. What is its volume?

(A) 36   (B) 12   (C) 60   (D) 180   (E) 216

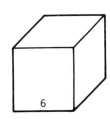

14. In the rectangular solid shown, $AB = 3$, $BC = 5$, and $BD = 8$. What is its volume?

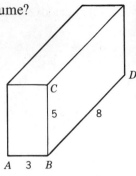

(A) 40   (B) 120   (C) 124   (D) 160   (E) 90

15. The right cylindrical shape shown to the right has a radius of 4 ft and a height of 9 ft. What is its volume?

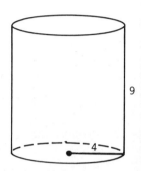

(A) 36π ft^3 (B) 81π ft^3 (C) 144π ft^3

(D) 45π ft^3 (E) 48 ft^3

16. The formula for finding the volume (V) of a sphere is: $V = \frac{4}{3}\pi r^3$. What is the volume of a sphere with radius 3?

(A) 36π (B) 24π (C) 27π (D) 108π (E) 108

17. In the rectangular solid shown to the right, $AB = 5$, $BC = 3$, and $CD = 4$. If all the dimensions are doubled, what will be the volume?

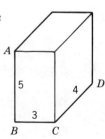

(A) 240   (B) 360   (C) 520   (D) 480   (E) 1440

18. If the radius of a sphere is doubled, how many times is the volume increased?

(A) twice   (B) 4 times   (C) 8 times   (D) $\frac{4}{3}$ times   (E) 2π times

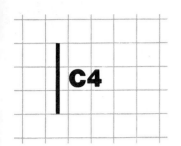

C4

THE SUM OF THE INTERIOR ANGLES OF A TRIANGLE

C4.1 TYPES OF TRIANGLES

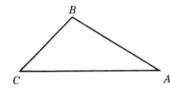

Here is a way to describe a **triangle**: three points in a plane, not all on the same line, along with the three line segments that join them.

Every triangle has three **angles**, three **vertices**, and three **sides**. The symbol for a triangle is "△", and the figure shows △ABC. The three *angles* in this triangle are ∠A, ∠B, and ∠C. The *vertices* are the points A, B, and C. The *sides* are AB, BC, and CA.

There are various ways to classify triangles. We list here some of the more common classifications. In these descriptions, the word **congruent** will arise. Two line segments are congruent if they have the same length. Two angles are congruent if they have the same degree measurement.

An **equilateral** triangle is a triangle with all three sides congruent. It is also true that all three angles of an equilateral triangle are congruent.

An **isosceles** triangle is a triangle with two sides that are congruent. Thus, every equilateral triangle is isosceles, but not every isosceles triangle is equilateral.

A **scalene** triangle is a triangle with no two sides congruent. This figure shows the three types of triangles that have now been presented. These three types have all been described by means of their sides. But triangles àre also described by means of their angles.

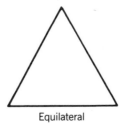

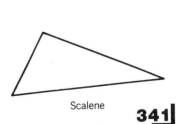

Equilateral   Isosceles   Scalene

An **acute** triangle is a triangle whose three angles are all acute (less than 90°).

A **right** triangle is a triangle with an angle of 90°. Right triangles will be the subject of a future unit.

An **obtuse** triangle is a triangle with an angle that is obtuse (greater than 90°).

This figure illustrates these concepts.

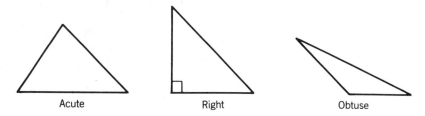

In any triangle, the terms **opposite** and **adjacent** will arise. In $\triangle ABC$ shown here, side AB is opposite $\angle C$, side BC is opposite $\angle A$, and side CA is opposite $\angle B$. The sides that are not opposite an angle are called adjacent to the angle. Thus, in a triangle, each angle has one opposite side and two adjacent sides.

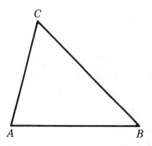

However, in a right triangle, one side has a special name: The side opposite the 90° angle is called the **hypotenuse** of the triangle.

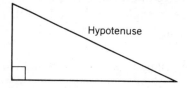

In labeling a triangle, corresponding small letters are usually (but not always!) used to name the sides that are opposite the angles or vertices named by capital letters.

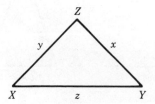

C4.1 PROBLEMS

Be sure to work *each* problem.

1. Use a ruler to draw a segment *CD* congruent to segment *AB*.

$\overline{}$
A   B

2. Use a protractor to draw an angle congruent to ∠*A*. (Don't worry about the lengths of the line segments in this figure.)

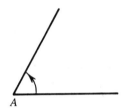

A

3. Use your protractor to draw a triangle with all three angles measuring 60°. Now measure the sides. What do you notice?

4. Draw a right triangle *ABC* with the 90° angle at *A*, and with sides *AB* and *AC* each of length 2 in. Now measure both ∠*B* and ∠*C*. How many degrees is each?

5. Draw any triangle ABC you wish, preferably a scalene triangle to make things interesting. Now draw $\triangle DEF$ so that $\angle A$ is congruent to $\angle D$, $\angle B$ is congruent to $\angle E$, $\angle C$ is congruent to $\angle F$, and DE is twice as long as AB. Is EF twice as long as BC? Is DF twice as long as AC?

6. Draw $\triangle ABC$ so that $\angle A$ is a right angle, $\angle B$ is $30°$ and $\angle C$ is $60°$. Thus, $\angle C$ is twice the size of $\angle B$. Is side c twice as long as side b?

7. In Problem 6, $\angle A$ is three times the measure of $\angle B$. Is the hypotenuse three times as long as side b? Is it two times as long?

8. Repeat Problem 7 several times, making a new triangle each time. Do you notice anything?

9. Draw any scalene triangle ABC. Now draw $\triangle DEF$ so that each side is twice as long as the corresponding sides of $\triangle ABC$. What do you notice about the angles of $\triangle DEF$ as compared with those of $\triangle ABC$?

C4.2 SUM OF INTERIOR ANGLES

Casey Stengel once said, "You can notice a lot just by looking." (He was famous for such deep philosophical insights.) We will use the Stengel approach in this unit.

Unlike the other units, you will here be asked to perform three experiments to see what you can discover. Perform each experiment at least three times to let experimental error "average out."

EXPERIMENT 1
Draw a segment AB of any length you wish (and use a different length each time). At A, draw an acute angle, and find its measure in degrees. Using your protractor, at B, draw an angle whose measure, added to that of $\angle A$, will sum to $90°$. For example, if $\angle A$ measures $42°$, then $\angle B$ must measure $48°$ since $48° + 42° = 90°$. Now extend the other sides of those two angles until they meet at C. Measure how many degrees $\angle C$ has. What do you notice each time?

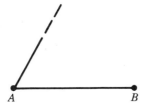

EXPERIMENT 2
Draw any triangles you wish. (Make at least one obtuse triangle for this experiment.) With a protractor, find the number of degrees in each. Find their sum. What do you notice each time?

EXPERIMENT 3
Take a sheet of paper and draw a triangle—any shape you wish, but don't make it acute each time. With scissors, very carefully cut it out. Now label each angle on the inside.

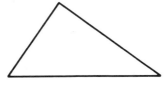

Next, tear off (don't use scissors) two of those angles.

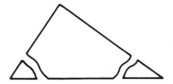

345

Finally, piece the two parts next to the untorn angle as shown.

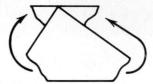

Make the sides fit together neatly. What do you notice each time?

From these three experiments, you have probably guessed (or concluded) that the sum of the angles in any triangle is very close to 180°. In fact, it can be proved that the sum of the angles in any triangle is exactly 180°. The mathematical proof of this fact emulates in every essential detail what was done in Experiment 3.

Now it's time for some problems.

C4.2 PROBLEMS

Measure or solve.

1. In $\triangle ABC$, what is the measure of $\angle B$?

   $\angle B =$ _____

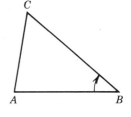

2. In $\triangle DEF$, what is the measure of $\angle D$?

   $\angle D =$ _____

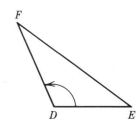

3. In this triangle, what is x?

   $x =$ _____

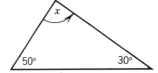

4. In this triangle, what is y?

   $y =$ _____

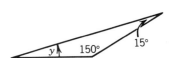

5. In $\triangle ABC$, $\angle B$ is 30° more than $\angle A$, and $\angle C$ is 40° more than $\angle B$. What is the size of each angle?

   $\angle A =$ _____ $\angle B =$ _____ $\angle C =$ _____

6. In $\triangle XYZ$, $\angle Y$ is 20° less than $\angle X$, and $\angle Z$ is three times the size of $\angle Y$. Give the size of each angle.

   $\angle X =$ _____ $\angle Y =$ _____ $\angle Z =$ _____

7. In △*PQR* shown here, what is the size of ∠*PQR*?

   ∠*PQR* = _____

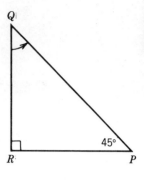

8. In △*GHI* as shown, what is the size of ∠*GIH*?

   ∠*GIH* = _____

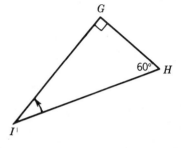

9. In an equilateral triangle, can all three angles be less than 60°? Explain.

10. In a right triangle, can either of the other angles be 90°? Explain.

11. In a quadrilateral (four-sided figure) *ABCD*, what is the sum of the angles? (*Hint*: Draw *AC* and use what you know about the sum of the angles in any triangle.)

    Sum of angles = _____

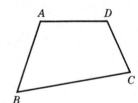

12. In a pentagon (five-sided figure) *EFGHI*, what is the sum of the angles? (*Hint*: Draw *FH* and *EH* and use what you know about the sum of the angles in any triangle.)

    Sum of angles = _____

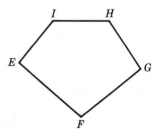

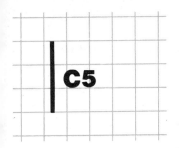

C5

PROPERTIES OF ISOSCELES AND EQUILATERAL TRIANGLES

C5.1 ISOSCELES TRIANGLES

An isosceles triangle is a triangle with two sides that have the same length. Study the following isosceles triangles.

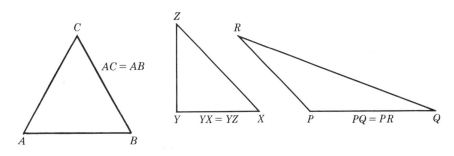

The angles that are opposite the sides of equal lengths in an isosceles triangle are called the **base angles**. The base angles in triangle ABC are $\angle A$ and $\angle B$ since they are opposite the two sides of equal lengths. Can you identify the base angles for triangles XYZ and PQR?

In triangle XYZ, $\angle X$ and $\angle Z$ are base angles. In triangle PQR, $\angle Q$ and $\angle R$ are base angles. It is also true that the base angles in an isosceles triangle have the same measure (number of degrees).

Example A   In the isosceles triangle ABC below, the two base angles, $\angle B$ and $\angle A$, each measures 30°.

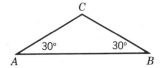

349

Example B In the isosceles triangle XYZ, $YX = YZ$. If $\angle X$ equals 20°, what is the degree measure of $\angle Z$? $\angle Z = 20°$

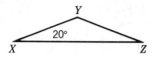

Example C In the isosceles triangle PQR to the right, $\angle P = \angle R$. What can be said about sides PQ and RQ?

Since they are the sides opposite equal angles in an isosceles triangle, they are of equal length.

The sum of the angles in a triangle is 180°, and we can use this fact when finding angle measures in isosceles triangles.

Example D In triangle XYZ, the length of XY is equal to the length of XZ. What is the degree measure of $\angle X$?

$\angle Z = 70°$. $\angle Y$ is the other base angle.

Add the two base angles $\angle Y + \angle Z = 140°$

Subtract this sum from 180° $180° - 140° = 40°.$

Therefore, $\angle X = 40°.$

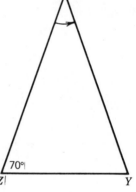

Example E In the isosceles triangle ABC to the right, we are given $AB = AC$ and $\angle A = 40°$. What do the other two angles measure in degrees?

First, we subtract $180° - 40° = 140°.$
   Then, since the two base angles in an isosceles triangle are equal, we can divide by 2.

$140° \div 2 = 70°$ Each angle $= 70°.$

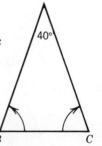

Information about the relationships of sides and base angles in isosceles triangles is useful in solving various problems in geometry.

C5.1 PROBLEMS

Find the degree measurement for each angle in the isosceles triangles below.

1. $\angle A =$ _____50°_____ $AC = BC$

   $\angle B =$ _____

   $\angle C =$ _____

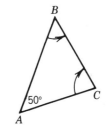

2. $\angle X =$ _____140°_____ $XY = XZ$

   $\angle Y =$ _____

   $\angle Z =$ _____

3. $\angle P =$ _____25°_____ $QR = RP$

   $\angle Q =$ _____

   $\angle R =$ _____

4. $\angle M =$ _____55°_____ $ON = OM$

   $\angle N =$ _____

   $\angle O =$ _____

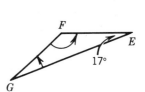

5. $\angle E =$ _____17°_____ $GF = FE$

   $\angle F =$ _____

   $\angle G =$ _____

351

6. $\angle H =$ ___65°___ $HI = IJ$

$\angle I =$ _____

$\angle J =$ _____

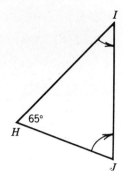

7. $\angle L =$ ___50°___ $MN = ML$

$\angle M =$ _____

$\angle N =$ _____

8. $\angle D =$ ___135°___ $DE = DF$

$\angle E =$ _____

$\angle F =$ _____

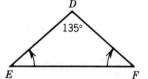

9. $\angle W =$ ___35°___ $YW = YX$

$\angle X =$ _____

$\angle Y =$ _____

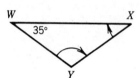

10. $\angle I =$ ___15°___ $KI = KJ$

$\angle J =$ _____

$\angle K =$ _____

C5.2 EQUILATERAL TRIANGLES

An equilateral triangle is a triangle with all three sides having the same length and all three angles having the same degree measure. These features of the equilateral triangle make it useful in many applications. Here is an equilateral triangle.

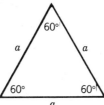

Example A In equilateral triangle ABC, $AB = 5$ units. What are the lengths of AC and BC?

Since all sides have the same length, and we know $AB = 5$ units, then $AC = 5$ units and $BC = 5$ units.

Example B In equilateral triangle XYZ, what is the angle measure for each angle?

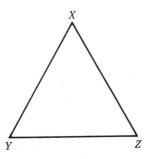

Since all angles have the same measure in an equilateral triangle, and the sum of the angle measures in every triangle is $180°$, we simply divide by 3.

$180° \div 3 = 60°$. $\angle X$, $\angle Y$, $\angle Z$ each measure $60°$.

Example C What is the sum of the interior angles of a rectangular hexagon formed from six equilateral triangles?

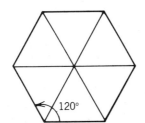

Since two angles of each equilateral triangle comprise one interior angle of the hexagon, we have $2 \times 60°$ or $120°$ in one interior angle.

There are six equal interior angles in the hexagon, thus the sum is: $6 \times 120°$ = $720°$.

Example D   If a regular hexagon has a perimeter of 12 units, what is the perimeter of each of the six equilateral triangles that make up the hexagon?

Since a regular hexagon has six equal sides and the total perimeter is 12 units, each side is $12 \div 6 = 2$ units.

Each equilateral triangle has three equal sides, one of which is 2 units. The total, then, is 3×2 or 6 units.

Example E   What is the sum of the angles of the six equilateral triangles that surround the center point of the hexagon in Example D?

Since each angle of an equilateral triangle is $60°$, and there are six of them, we have: $6 \times 60° = 360°$. This is, of course, the number of degrees about any point that is the center of a circle.

Sometimes we can combine information we know about triangles to solve more complicated problems.

Example F   In the figure shown, XY is a line segment, and $XZ = ZY = YP = PZ$. What is the number of degrees in $\angle X$?

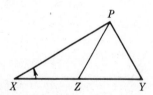

We know that XY is a line segment and that $\angle PZY$ is $60°$ since it is an angle of equilateral triangle ZYP. Thus, $\angle XZP = 180° - 60° = 120°$.

Since $XZ = ZP$, triangle XZP is an isosceles triangle. That means $\angle X = \angle XPZ$.

There are $180°$ in any triangle and $\angle XZP$ is $120°$, which means there are $60°$ to be equally divided between $\angle X$ and $\angle XPZ$. $60° \div 2 = 30°$. Thus $\angle X = 30°$, as does $\angle XPZ$.

Equilateral triangles are one of the few shapes that will fit together to cover a surface completely. This is why they are often seen in floor tile patterns.

C5.2 PROBLEMS

Determine the total perimeter for each of the following regular hexagons from the information given in the diagram.

1. Perimeter = _____

2. Perimeter = _____

3. Perimeter = _____

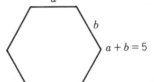

4. Perimeter = _____

The following shapes were constructed using isosceles or equilateral triangles. Sketch them in, using the fewest number of such triangles possible, and then determine the sum of the interior angles for each shape.

5. Number of triangles: _____

   Sum of the interior angle degrees: _____

6. Number of triangles: _____

   Sum of interior angle degrees: _____

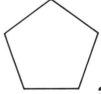

355

7. Number of triangles: _____

   Sum of interior angle degrees: _____

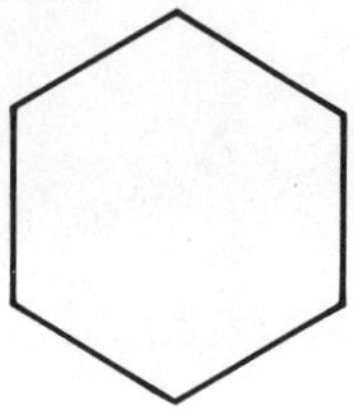

8. Number of triangles: _____

   Sum of interior angle degrees: _____

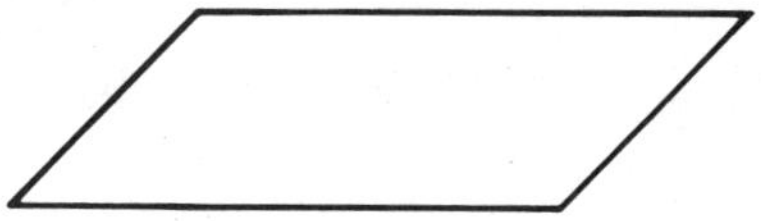

9. Number of triangles: _____

   Sum of interior angle degrees: _____

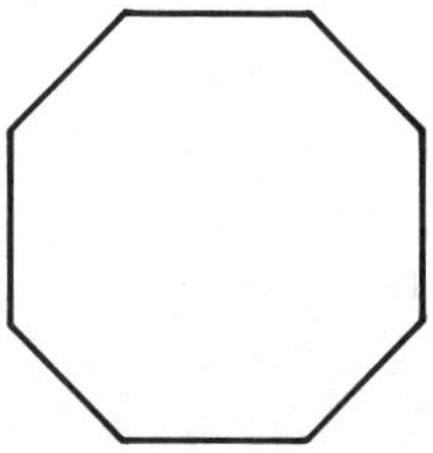

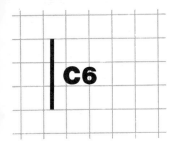

C6

PROPERTIES OF SIMILAR AND CONGRUENT TRIANGLES

C6.1 CONGRUENT TRIANGLES

The concept of **congruence** appears often in geometry. It simply means that two, or more, geometric figures have the same size and shape but not the same location. One easy way to determine if two geometric shapes are congruent is to place one on top of the other and see if they fit. If sides and angles fit exactly, then the shapes are congruent. In general, two polygons are congruent if they have corresponding sides and angles that are congruent.

A triangle is made up of three angles and three sides. Two triangles are congruent when their three pairs of corresponding angles and their three pairs of corresponding sides are congruent. This means they will fit on top of one another perfectly. We use a special symbol for congruent: " $\cong$ ".

Example A   Show why triangle ABC is congruent to triangle XYZ.

Angles	Sides
$\angle A \cong \angle X$	$AB \cong XY$
$\angle B \cong \angle Y$	$BC \cong YZ$
$\angle C \cong \angle Z$	$CA \cong ZX$

Corresponding parts are congruent; therefore, the two triangles are congruent.

The question now arises: do we have to have all this information about two triangles to determine if they are congruent or not? The answer is no, but be careful because certain conditions must be met.

357

Example B If the three pairs of corresponding sides in two triangles are congruent, are the triangles congruent?

Yes. It follows that the angles are also congruent.

This relationship is sometimes called **SSS**: Side, Side, Side

$AB \cong A'B'$

$BC \cong B'C'$

$CA \cong C'A'$

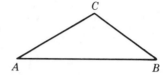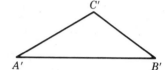

Example C If we know that two sides and the angle between them are congruent in two triangles, are the triangles congruent?

The answer is yes.

This is sometimes remembered as **SAS**: Side, Angle, Side. The angle must be the included angle.

$PR \cong P'R'$

$PQ \cong P'Q'$

$\angle P \cong \angle P'$

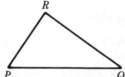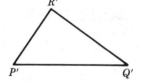

Example D If we know that two angles and the side between them are congruent in two triangles, are the triangles congruent?

The answer is yes.

This may be remembered as **ASA**, or Angle, Side, Angle. The side must be the included side.

$\angle M \cong \angle M'$

$\angle N \cong \angle N'$

$MN \cong M'N'$

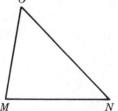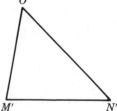

Example E Another condition that results in congruence is when two angles and the side opposite one of the angles in one triangle are congruent to the corresponding parts in the other triangle. This is referred to as **AAS**: Angle, Angle, Side.

$\angle X \cong \angle A$

$\angle Y \cong \angle B$

$YZ \cong BC$

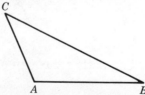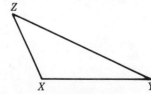

You can easily convince yourself that the above conditions between two triangles will result in their being congruent by doing some constructions (drawings). Formal proofs can also be done.

C6.1 PROBLEMS

In each of the following, you are given information about pairs of triangles. Determine if this information makes the two triangles congruent.

1. Given:

     $AB \cong XY$

     $BC \cong YZ$

     $CA \cong ZX$

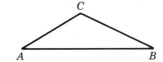 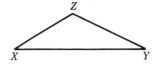

Is triangle ABC congruent to triangle XYZ? _____ Why? _____

2. Given:

     $\angle P \cong \angle U$

     $PQ \cong UV$

     $\angle Q \cong \angle V$

Is triangle PQR congruent to triangle UVW? _____ Why? _____

3. Given:

     $MN \cong PQ$

     $MO \cong PR$

     $\angle M \cong \angle P$

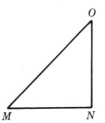 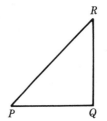

Is triangle MNO congruent to triangle PQR? _____ Why? _____

4. Given:

     $\angle R \cong \angle D$

     $\angle S \cong \angle E$

     $RS \cong DE$

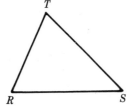 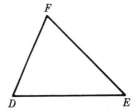

Is triangle RST congruent to triangle DEF? _____ Why? _____

359

5. Given:

$\angle O \cong \angle Z$

$\angle N \cong \angle Y$

$MO \cong XZ$

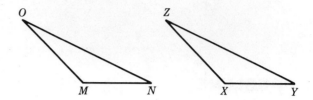

Is triangle *MNO* congruent to triangle *XYZ*? _____ Why? _____

6. Given:

$AB \cong AE$

$CA \cong AD$

$\angle BAC \cong \angle EAD$

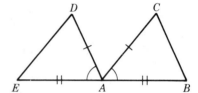

Is triangle *ABC* congruent to triangle *AED*? _____ Why? _____

C6.2 SIMILAR TRIANGLES

Two triangles are **similar** if the three angles of the first triangle are congruent to the three corresponding angles of the other. The "size" of the triangles has nothing to do with their being similar—similarity pertains only to the angles.

Example A Triangle ABC is similar to triangle XYZ because their corresponding angles are congruent. $\angle A \cong \angle X, \angle B \cong \angle Y, \angle C \cong \angle Z.$

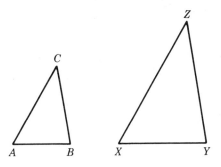

Since the sum of the angles in every triangle is $180°$, it follows that, if in two triangles two pairs of corresponding angles are congruent, the third pair must also be congruent.

Example B Given two triangles, ABC and XYZ, and $A = 30°$, $X = 30°$, $B = 70°$, $Y = 70°$, $\angle C$ is congruent to $\angle Z$ because in

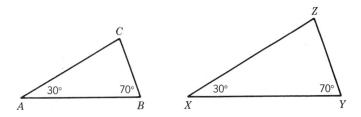

triangle ABC: $180 - (30 + 70) = 80$, and in

triangle XYZ: $180 - (30 + 70) = 80$.

When two triangles are similar, a ratio is established between corresponding parts of the two triangles. These ratios are very useful in solving problems.

Example C Similar triangles ABC and XYZ, shown on the right, are constructed to meet the following conditions:

$$\frac{AB}{XY} = \frac{BC}{YZ} = \frac{CA}{ZX} = \frac{2}{3}$$

From this it follows that

$\angle A \cong \angle X; \angle B \cong \angle Y; \angle C \cong \angle Z.$

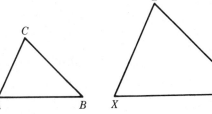

When a ratio is established in similar triangles, it can be used to solve for unknown values by using a proportion (a statement of equality between two ratios).

To solve a proportion, cross multiply and solve for the unknown quantity.

$$\frac{1}{2} = \frac{x}{8} \qquad 2x = 8,$$

thus

$$x = 4$$

Applying this technique to similar triangles, when given three values you can solve for the fourth.

Example D   Given similar right triangles PQR and UVZ, with values indicated, find the length of RQ.

$$\frac{3}{8} = \frac{x}{24} \quad \text{or} \quad 8x = 72 \qquad x = 9$$

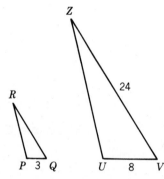

Example E   In the similar triangles to the right, $\angle ABD \cong \angle DCE$.

What is the length of AE?

Here the proportion becomes:

$$\frac{AD}{AB} = \frac{DE}{CD} \quad \text{or} \quad \frac{5}{2} = \frac{x}{6}$$

$$2x = 30$$

$$x = 15$$

So $AE = 5 + 15 = 20$.

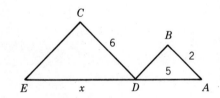

Example F   In the following figure, find the length of AE.

$ED\|CB$.   (‖ means parallel)

$$\frac{5}{x + 2} = \frac{3}{x}$$

thus

$$5x = 3x + 6 \quad \text{and} \quad 2x = 6 \qquad x = 3$$

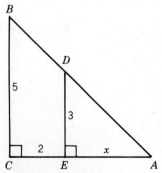

Applications for using the properties of similar triangles are frequent.

C6.2 PROBLEMS

Triangles *ABC* and *XYZ* are similar. Answer questions 1–5.

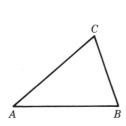 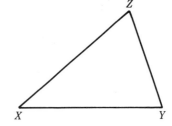

1. List the pairs of angles that are congruent.

   a.

   b.

   c.

2. Complete the following: $\dfrac{AC}{AB} = \dfrac{XZ}{\underline{}}$; $\dfrac{ZY}{CB} = \dfrac{\underline{}}{AC}$.

3. If $AC = 6$ units, $AB = 8$ units, $XZ = 12$ units, what is the length of XY?

4. If $\angle A = 60°$, $\angle B = 40°$, how many degrees in $\angle Z$?

5. If $ZY = 15$, $XY = 18$, $CB = 8$, what is the length of AB?

6. In the figure shown, AE is parallel to BD. What is the length of AB?

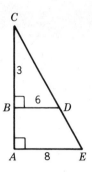

7. In the figure shown, TP is parallel to SQ. What is the length of QR?

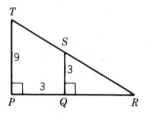

8. A 6-ft stick casts a shadow of 10 ft. If a tree casts a shadow of 40 ft at the same time, what is the height of the tree?

9. In the figure below, what is the ratio of XB to YC? $XY \| BC$?

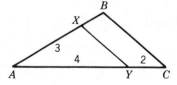

C4	**ACHIEVEMENT TEST** THE SUM OF THE INTERIOR ANGLES OF A TRIANGLE
C5	PROPERTIES OF ISOSCELES AND EQUILATERAL TRIANGLES
C6	PROPERTIES OF SIMILAR AND CONGRUENT TRIANGLES

1. The sum of the angles in a right triangle is

   (A) 160   (B) Varies   (C) 175   (D) 180   (E) 200

2. In the isosceles triangle shown to the right, $XY = ZY$ and $\angle XYZ = 30°$. What is the degree measure of $\angle X$?

   (A) 75   (B) 60   (C) 55   (D) 80   (E) Can't tell

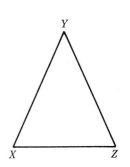

3. In triangle XYZ shown to the right, the degree measure of $\angle X$ is

   (A) 110   (B) 120   (C) 130   (D) 140   (E) 95

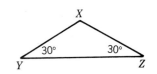

4. In the figure shown to the right, segment AB is perpendicular to segment CD. What is the measure of $\angle BDC$?

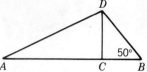

   (A) 60   (B) 50   (C) 45   (D) 40   (E) 30

5. A triangle where all angles have the same measure is called

   (A) Golden   (B) Right   (C) Scalene   (D) Equilateral

   (E) Isosceles

6. A triangle with two angles having equal measures is called

   (A) Scalene   (B) Isosceles   (C) Golden   (D) Equilateral

   (E) Right

7. The figure to the right is made of three equilateral triangles. What is the sum of the interior angles of the figure $ABCD$?

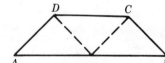

   (A) 360   (B) 240   (C) 320   (D) 180   (E) 320

8. In the figure to the right, $\triangle XYZ$ is equilateral and Z is the midpoint of segment XW. What is the measure of $\angle W$?

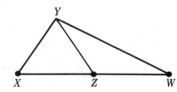

   (A) 30   (B) 40   (C) 45   (D) 20   (E) 60

9. In the figure to the right, triangle *PQR* is isosceles. *PQ = RQ* and ∠*QRS* = 110°. What is the degree measure of ∠*PQR*?

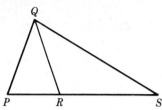

(A) 30   (B) 40   (C) 50   (D) 60   (E) 70

10. Six equilateral triangles form the regular hexagon shown to the right. If the perimeter of each triangle is $\sqrt{5}$, what is the perimeter of the hexagon?

(A) $3\sqrt{5}$ (B) $6\sqrt{5}$ (C) $\dfrac{\sqrt{5}}{3}$ (D) $\dfrac{\sqrt{5}}{2}$ (E) $2\sqrt{5}$

11. In the figure to the right, what is the degree measure of ∠*DCA*?

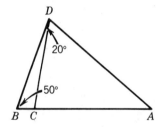

(A) 60   (B) 70   (C) 80   (D) 50   (E) 65

12. In the diagram to the right, *XZ∥VW*. What is the length of *XY*?

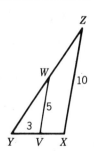

(A) 3   (B) 4   (C) 5   (D) 6   (E) 7

13. In the two triangles *ABC* and *XYZ*, ∠*A* = ∠*X* and ∠*B* = ∠*Y*. For the two triangles to be congruent, which of the following must also be true?

(A) ∠*A* = ∠*Z*   (B) *BC = XY*   (C) *AC = YZ*   (D) *AB = XY*

(E) *AB = XZ*

14. In the figure to the right, $AB = AC$ and $DE \parallel BC$. What is the length of DE?

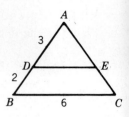

(A) 2   (B) 3   (C) 4   (D) 4.5   (E) 3.6

15. A tree casts a shadow 60 ft long while a 6-ft man casts a shadow of 10 ft. How tall is the tree?

(A) 30 ft   (B) 36 ft   (C) 48 ft   (D) 60 ft   (E) 72 ft

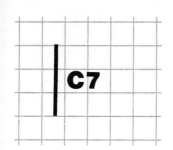

C7

THE PYTHAGOREAN THEOREM AND SPECIAL TRIANGLES

C7.1 THE PYTHAGOREAN THEOREM

In the right triangle of Figure C7-1, c is the length of the hypotenuse and a and b are the lengths of the legs.

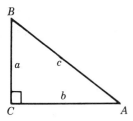

Figure C7-1

The following relationships of these three lengths to each other is called The **Pythagorean theorem**. It is one of the most useful, important, and well known theorems in all of mathematics:

$$c^2 = a^2 + b^2$$

Example A Using your ruler, measure the length of a, b, and c in the right triangle in Figure C7-1. Verify this relationship.

You should have found the lengths to be as follows (one unit is $\frac{3}{4}$ centimeters):

$a = 3$ units

$b = 4$ units

$c = 5$ units

Does $5^2 = 3^2 + 4^2$? Yes, since $5^2 = 25$, $3^2 = 9$, $4^2 = 16$, and $25 = 9 + 16$.

When three positive numbers are so related that the sum of the squares of two of them is equal to the square of the third, the set of the three numbers is called a **Pythagorean triple**.

Example B   3, 4, 5 is a Pythagorean triple. If each of these numbers is doubled, we get 6, 8, 10. Is 6, 8, 10 a Pythagorean triple?

Yes. A triangle with sides of 6, 8, and 10 is similar to a triangle with sides of 3, 4, and 5 because the corresponding sides of the 2 triangles are in proportion. Since the triangle with sides of 3, 4, and 5 is a right triangle, the triangle with sides of 6, 8, and 10 is also a right triangle. So 6, 8, 10 is another Pythagorean triple. Check: $10^2 = 6^2 + 8^2$ since $100 = 36 + 64$.

The Pythagorean theorem for any triangle similar to the triangle in Figure C7-1 can be verified geometrically by drawing a square on each side of the right triangle to represent the square of the length of that side. By dividing each square into 1×1 square units as shown in Figure C7-2, the Pythagorean relationship is seen.

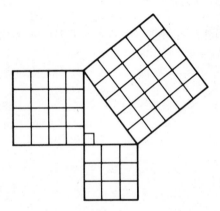

Figure C7-2

Count the 1×1 square units in Figure C7-2. The number of units in the square on side c equals the sum of the numbers of units in the squares on sides a and b.

A way to verify the Pythagorean theorem for any right triangle is to draw a square on each side of a right triangle, then cut the squares on the two legs to fit exactly into the square on the hypotenuse. Figure C7-3 illustrates one way to do this.

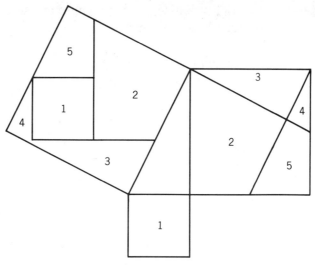

Figure C7-3

The Pythagorean theorem can be stated in words:

In a right triangle, the square of the length of the hypotenuse is equal to the sum of the squares of the lengths of the legs.

Example C In the right triangle shown in Figure C7-4, the legs measure 5 and 12. Find the hypotenuse.

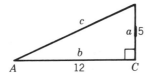

Figure C7-4

Applying the Pythagorean theorem:

$$c^2 = 5^2 + 12^2$$

$$c^2 = 25 + 144$$

$$c^2 = 169$$

$$c = \sqrt{169} = 13$$

The Pythagorean theorem can also be used to find the length of one leg of a right triangle if the lengths of the other two sides are known.

Example D   In a right triangle, the hypotenuse measures 6, and one leg measures 5. What is the length of the other leg?

By the Pythagorean theorem,

$$a^2 + 5^2 = 6^2$$
$$a^2 + 25 = 36$$
$$a^2 = 11$$
$$a = \sqrt{11}$$

C7.1 PROBLEMS

In Problems 1–5, verify that the lengths given satisfy $c^2 = a^2 + b^2$ by filling in the blanks.

	c	a	b	c^2	a^2	b^2
1.	17	8	15	___	___	___
2.	6.25	6	1.75	___	___	___
3.	13	5	12	___	___	___
4.	3	2	$\sqrt{5}$	___	___	___
5.	25	20	15	___	___	___

Complete the following table for Problems 6–18. If an answer is not an exact square root, use the square root sign or estimate it to the nearest tenth.

	Leg a	Leg b	Hypotenuse c
6.	___	8	10
7.	9	12	___
8.	___	$3\sqrt{6}$	8
9.	3	___	$\sqrt{15}$
10.	5	5	___
11.	$\sqrt{10}$	$\sqrt{15}$	___
12.	___	1.5	2.5
13.	5	4	___
14.	24	26	___
15.	5	1	___
16.	1	$1\frac{1}{2}$	___
17.	3.6	4.8	___
18.	7.5	___	19.5

C7.2 SPECIAL TRIANGLES

The 45°, 45°, 90° triangle is **special**. It is an isosceles right triangle and the legths of the sides have special relationships.

In triangle ABC, $AC = BC$. The length of the hypothenuse, or longest side, is the product of the length of one of the equal sides and $\sqrt{2}$. That is, $AB = AC\sqrt{2}$ and $AB = BC\sqrt{2}$.

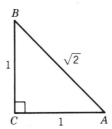

This can be verified by applying the Pythagorean theorem. In the triangle above, let $AC = 1$ unit. Then $BC = 1$ unit.

$$(AB)^2 = 1^2 + 1^2 \quad \text{or} \quad 2$$

$$AB = \sqrt{2}$$

It can also be shown that, when given the length of the hypotenuse, the length of each of the equal sides in a 45°, 45°, 90° triangle can be found by dividing the length of the hypotenuse by $\sqrt{2}$.

That is, in triangle ABC, $AC = \dfrac{AB}{\sqrt{2}}$ and $BC = \dfrac{AB}{\sqrt{2}}$.

This information is useful: When the length of one side of a 45°, 45°, 90° triangle is known, the other two lengths are found easily.

Example A   Given triangle ABC as shown, find BC and AB.

$AC = 5$ units

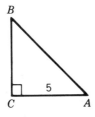

Because triangle ABC is isosceles, $AC = BC$. Thus, $BC = 5$ units. Because triangle ABC is an isosceles right triangle, the length of the hypotenuse, AB, is the product of the length of one of the equal sides and $\sqrt{2}$. Thus, $AB = 5\sqrt{2}$.

375

Example B   Given isosceles right triangle XYZ as shown, find XZ and YZ.

$XY = 8$ units

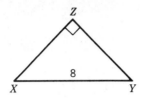

Because triangle XYZ is an isosceles right triangle, the length of each of the equal sides is the length of the hypotenuse divided by $\sqrt{2}$. (Sometimes it is necessary to "rationalize" the denominator as shown.)

$$XZ = \frac{8}{\sqrt{2}} = \frac{\sqrt{2}\,(8)}{\sqrt{2}\,(\sqrt{2})} = \frac{8\sqrt{2}}{2} = 4\sqrt{2} \quad \text{and} \quad YZ = 4\sqrt{2}$$

The diagonal of a square divides it into two triangles with angles measuring $45°$, $45°$, $90°$.

Example C   In the figure shown, square $MNOP$ has a perimeter of 36 units. Find the length of the diagonal MO.

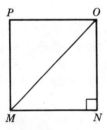

One side of a square with a perimeter of 36 units has a length of $36 \div 4$ or 9 units.

The diagonal MO is the hypotenuse of a $45°$, $45°$, $90°$ triangle. Its length is the product of the length of one of the equal sides and $\sqrt{2}$.

$MO = 9\sqrt{2}$ units

Another special triangle has angles that measure $30°$, $60°$, $90°$.

Again, the lengths of the sides of this right triangle have special relationships.

The length of the hypotenuse is twice the length of the side opposite the angle measuring $30°$, and the length of the side opposite the angle with a measurement of $30°$ is one-half the length of the hypotenuse.

In triangle ABC, as shown, these relationships are:

$$AB = 2AC \quad \text{and} \quad AC = \frac{1}{2}AB$$

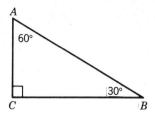

The lengths of the sides AC and BC have these relationships:

$$BC = AC\sqrt{3} \quad \text{and} \quad AC = \frac{BC}{\sqrt{3}}$$

These relationships can be verified by the Pythagorean theorem. But knowing the relationships can save you time when working problems that involve finding the lengths of the sides of a 30°, 60°, 90° triangle.

Example D   Given triangle ABC as shown, find AB and AC.

$BC = 6$ units

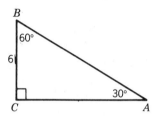

Because this is a 30°, 60°, 90° triangle, the length of the hypotenuse is twice the length of the side opposite the angle measuring 30°. Since $BC = 6$ units, then $AB = 12$ units.

The length of the side opposite the angle measuring 60° is the product of the length of the side opposite the angle measuring 30° and $\sqrt{3}$.
Since $AB = 6$ units, then $AC = 6\sqrt{3}$.

Example E   Given triangle XYZ, as shown, find XZ and YZ.

$XY = 10$ units

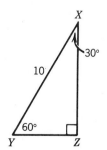

Because this is a 30°, 60°, 90° triangle, the length of the side opposite the angle measuring 30° is one-half the length of the hypotenuse.

Since $XY = 10$ units, then $YZ = 5$ units.

The length of the side opposite the angle measuring 60° is the product of the length of the side opposite the angle measuring 30° and $\sqrt{3}$.

Since $YZ = 5$ units, then $XZ = 5\sqrt{3}$ units.

Example F   Given the figure shown, find AB, AC, CD, BD.

$BC = 15$ units

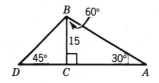

In triangle ABC: In triangle BCD:

$AB = 2\,BC = (2)(15) = 30$ units $CD = BC = 15$ units

$AC = \sqrt{3}\,BC = 15\sqrt{3}$ units $BD = \sqrt{2}\,BC = 15\sqrt{2}$ units

C7.2 PROBLEMS

Solve.

1. In right triangle ABC as shown, what is the length of AB?

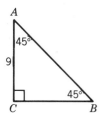

2. In right triangle XYZ as shown, what is the length of XZ?

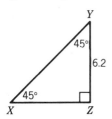

3. In right triangle DEF as shown, what is the length of EF?

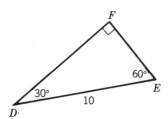

4. In right triangle JKL as shown, what is the length of JK?

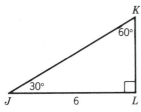

5. In square $ABCD$ as shown, what is the length of the diagonal?

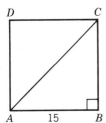

6. In square *RSTU* as shown, what is the perimeter?

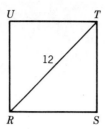

7. What is the ratio of the length of a diagonal of a square to its perimeter?

8. What is the ratio of the perimeter of a square to the length of a diagonal?

9. In the figure shown, what is the length of *DC*?

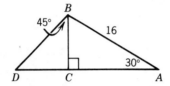

10. Using the figure in Problem 9, what is the length of *DB*?

11. Using the figure in Problem 9, what is the length of *AD*?

12. Using the information as shown, what is the degree measure of the smallest angle?

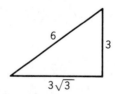

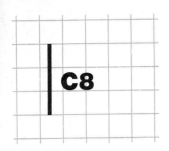

C8

PARALLELISM AND PERPENDICULARITY

C8.1 PARALLELISM

Lines that lie on the same plane are **parallel** if they have no points in common; that is, they will not intersect no matter how far they are extended. The symbol for "is parallel to" is "‖".

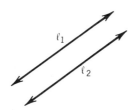

Line ℓ_1 and ℓ_2 are parallel to each other; $\ell_1 \| \ell_2$

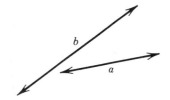

Lines a and b are not parallel to each other

There are many examples of parallel lines in the world. Parallelism is an extremely important concept.

When two parallel lines are intersected by a third line, called a **transversal**, many angles are formed.

Some knowledge of angles is useful in determining the relationships between the angles formed by two parallel lines intersected by a transversal.

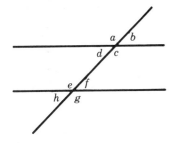

1. Two angles are supplementary if the sum of their measures is 180°.

In the preceding drawing, there are several pairs of supplementary angles. The angles in the following list are supplementary because the two angles together form a straight angle, a 180° angle.

a and b e and f
b and c f and g
c and d g and h
d and a h and e

381

2. Vertical angles have the same measure.

In the preceding drawing, these are the vertical angles that are formed.

a and c e and g
b and d f and h

3. Alternate interior angles have the same measure.

From the drawing, the alternate interior angles are listed.

c and e d and f

From this information, and knowing the measure in degrees of just one of these eight angles, you can find the measures of the remaining seven angles.

Example A   In the figure $\ell_1 \| \ell_2$ and $\angle w = 37°$. Find the measures of the other angles.

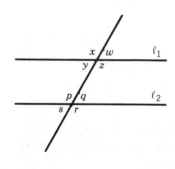

Since angle w and angle y are vertical angles, $\angle y$ measures $37°$.

Angle w is a supplementary angle to both angles x and z. That means the sum of angle w and x is $180°$, and the sum of angle w and angle z is $180°$.
Thus, angle x = angle z = $180° - 37° = 143°$.

Angles y and q are alternate interior angles and have the same measure.

Angles z and p are also alternate interior angles having the same measure.

$\angle y = \angle q = 37°$ $\angle z = \angle p = 143°$

Since angles q and s are vertical angles and angles p and r are vertical angles,

$\angle q = \angle s = 37°$ $\angle p = \angle r = 143°$

It can be seen easily that all of the angles in this problem measure one of two number of degrees, either $37°$ or $143°$.

All that must be known to find the measures of the angles formed by a transversal intersecting two parallel lines is the measure of one of the angles.

Example B   In the figure shown, find the measure of $\angle x$.

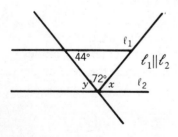

One way to solve this problem is to note that the measure of angle y is the same as the angle marked $44°$ because they are alternate interior angles.

Then, $\angle x + 72° + 44° = 180°$ because they form a straight angle.

$\angle x = 180° - 72° - 44° = 64°$

C8.1 PROBLEMS

1. Identify all of the pairs of lines that appear to be parallel.

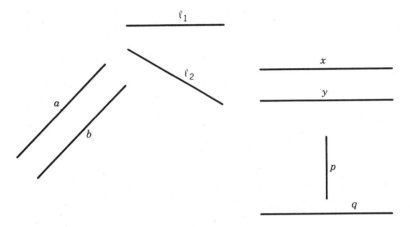

2. Draw three lines that are all parallel to each other.

3. In the figure shown, if $\ell_1 \| \ell_2$, name the pairs of vertical angles.

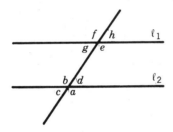

4. From the figure in Problem 3, name the pairs of alternate interior angles.

5. In the figure in Problem 3, if $\angle c$ measures $52°$, give the measurements of all of the other angles.

6. In the figure shown, if $\ell_1 \| \ell_2$, then how many degrees does $\angle x$ measure?

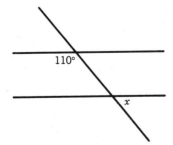

383

7. In the figure shown, if $\ell_1 \| \ell_2$, then what is the value of x?

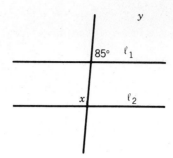

8. In the figure shown, if $\ell_1 \| \ell_2$, then what is the value of x?

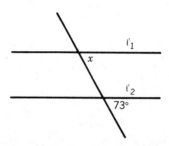

9. In the figure shown, $\ell_1 \| \ell_2$. Find the measure of $\angle x$.

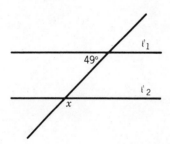

10. In the figure shown, $\ell_1 \| \ell_2$. Find the measure of $\angle y$, in terms of $\angle x$.

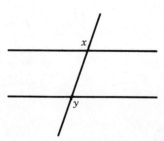

C8.2 PERPENDICULARITY

You have seen what it means for two lines to be parallel: They have the same *slope*. The slope of a line is determined by its relation to a horizontal line and is defined by

$$\text{Slope} = \frac{\text{Rise}}{\text{Run}}$$

as shown here.

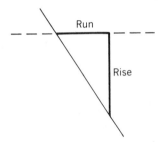

In the figure shown, the slope is positive since the rise is up (positive) and the run is to the right (positive). You can always make the run to the right, though, and so the slope really depends on the rise. In this next figure, the slope is negative.

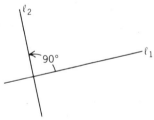

Two lines are perpendicular if they meet at right angles. The next figure shows perpendicular lines ℓ_1 and ℓ_2.

Let us now examine the slopes of ℓ_1 and ℓ_2 in two figures.

You see that the

$$\frac{\text{Rise of } \ell_1}{\text{Run of } \ell_1} = -\frac{\text{Run of } \ell_2}{\text{Rise of } \ell_2}$$

In other words,

$$\text{Slope of } \ell_1 = -\frac{1}{\text{Slope of } \ell_2}$$

or, put another way,

$$(\text{Slope of } \ell_1)(\text{Slope of } \ell_2) = -1$$

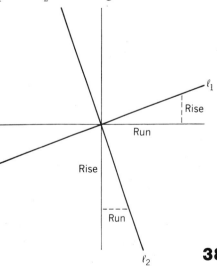

385

If this is true, we write $\ell_1 \perp \ell_2$. This leads to our first statement about perpendicular lines.

I. The slopes of perpendicular lines are negative reciprocals of one another.

(This does not hold, however, for a horizontal and a vertical line. They are, of course, perpendicular, but a vertical line has no slope because the run is 0.)

Example A The slope of ℓ_1 is $\frac{3}{4}$ and $\ell_1 \perp \ell_2$. What is the slope of ℓ_2?

Since $\ell_1 \perp \ell_2$, then the slope of ℓ_2 is $-\frac{4}{3}$, since $\left(\frac{3}{4}\right)\left(-\frac{4}{3}\right) = -1$.

There are two more statements that can be made about parallel and perpendicular lines.

II. If $\ell_1 \| \ell_2$ and $\ell_1 \perp \ell_3$, then $\ell_2 \perp \ell_3$.

This is true because ℓ_1 and ℓ_2 have the same slopes.

III. If $\ell_1 \perp \ell_3$ and $\ell_1 \| \ell_2$, then $\ell_2 \perp \ell_3$.

This is just another wording of II.

There is another way to determine perpendicular lines. It is through the converse of the Pythagorean theorem.

IV. If, in a triangle, the square of one side is equal to the sum of the squares of the other two sides, then the triangle is a right triangle.

Example B A triangle has sides of lengths 2, 3, and s. What must s be so that the two indicated sides are perpendicular?

By statement IV, s^2 must equal $2^2 + 3^2$:

$$s^2 = 2^2 + 3^2$$
$$= 4 + 9$$
$$= 13$$

Therefore,

$$s = \sqrt{13}$$

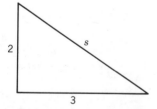

C8.2 PROBLEMS

Solve.

1. a. If $\ell_1 \| \ell_2$ and the slope of ℓ_1 is $\frac{2}{3}$, what is the slope of ℓ_2?

   b. If $\ell_1 \perp \ell_2$ and the slope of ℓ_1 is $\frac{2}{3}$, what is the slope of ℓ_2?

2. a. If $\ell_1 \perp \ell_2$ and the slope of ℓ_1 is $\frac{3}{5}$, what is the slope of ℓ_2?

   b. If $\ell_1 \| \ell_2$ and the slope of ℓ_1 is $\frac{3}{5}$, what is the slope of ℓ_2?

3. If $\ell_1 \| \ell_2$ and $\ell_2 \perp \ell_3$, what is the relationship between ℓ_1 and ℓ_3?

4. If $\ell_1 \perp \ell_2$ and $\ell_2 \| \ell_3$, what is the relationship between ℓ_1 and ℓ_3?

5. The sides of a triangle are of lengths 3, 5, and 6. Are two of the sides perpendicular?

6. The sides of a triangle are of lengths 6, 8, and 10. Are two of the sides perpendicular?

7. In the figure shown, what is x?

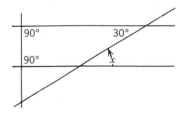

8. In the figure shown, what is x?

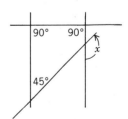

9. A triangle has sides of lengths 3, 8, and x. List all the possible values of x for which this could be a right triangle.

10. A triangle has sides of lengths 5, 7, and t. List all the possible values of t for which this could be a right triangle.

11. Lines ℓ_1, ℓ_2, and ℓ_3 are all parallel, and $\ell_4 \perp \ell_2$. The slope of ℓ_3 is $\frac{3}{8}$. What is the slope of ℓ_4?

12. Line $\ell_1 \perp \ell_2$, and $\ell_2 \perp \ell_3$, and $\ell_3 \perp \ell_4$. What is the relationship between ℓ_2 and ℓ_4?

13. In a right triangle, one of the angles is $37°$. Give a list of the sizes of all the angles.

14. In a right triangle, two of the angles are the same size. Give a list of the sizes of all the angles.

15. Show why a triangle with sides 5, 8, and 9 is not a right triangle.

16. Show why a triangle with sides 3, 5, and 6 is not a right triangle.

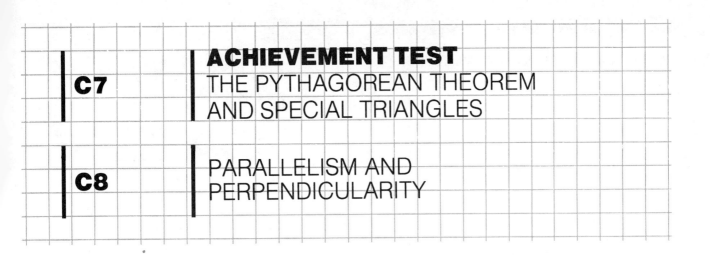

1. In right triangle ABC, what is the length of AB?

   (A) $\sqrt{21}$ (B) $\sqrt{29}$ (C) 29   (D) 7   (E) 4

2. In right triangle ABC, what is the length of BC?

   (A) $2\sqrt{5}$ (B) $\sqrt{10}$ (C) $2\sqrt{15}$ (D) 2   (E) 20

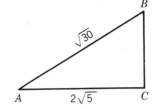

3. In right triangle XYZ, what is the length of XY?

   (A) $8\sqrt{5}$ (B) 8   (C) $16\sqrt{3}$ (D) $\frac{8}{3}\sqrt{3}$ (E) $8\sqrt{3}$

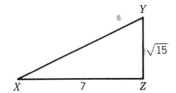

4. In right triangle XYZ, what is the length of XZ?

   (A) 7   (B) $7\sqrt{2}$ (C) $\frac{7}{2}\sqrt{2}$ (D) $7\sqrt{3}$ (E) 14

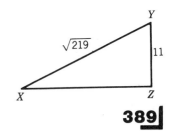

5. In triangle PQR, what is the length of PR?

   (A) $6\sqrt{2}$ (B) $3\sqrt{2}$ (C) 8   (D) $4\sqrt{2}$ (E) $2\sqrt{6}$

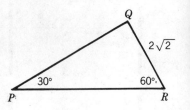

6. In triangle PQR, what is the length of PR?

   (A) 7   (B) $14\sqrt{3}$ (C) $7\sqrt{3}$ (D) $\dfrac{7}{3}\sqrt{3}$ (E) $7\sqrt{2}$

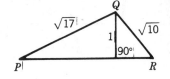

7. In the right triangle BCD, what is the length of BD?

   (A) $9\sqrt{3}$ (B) $\dfrac{9}{2}\sqrt{2}$ (C) $9\sqrt{2}$ (D) 9   (E) 18

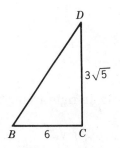

8. In the figure shown, if $\ell_1 \| \ell_2$, then $y =$

   (A) $108°$ (B) $96°$ (C) $84°$ (D) $72°$ (E) $24°$

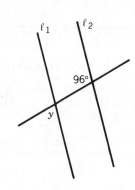

9. In the figure shown to the right, if $\ell_1 \perp \ell_2$, then $x =$

   (A) $120°$ (B) $15°$ (C) $105°$ (D) $75°$ (E) $165°$

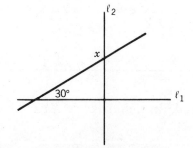

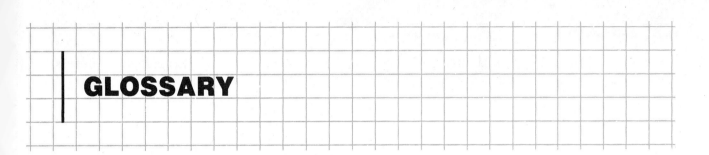

GLOSSARY

Abscissa The first element of an ordered pair, (x, y), denoting the horizontal coordinate. See *Ordinate*.

Absolute value The absolute value of a number is the distance the number is from zero. That is, the absolute value of n, written $|n| = \begin{cases} n & \text{if } n \geq 0 \\ -n & \text{if } n < 0 \end{cases}$.

Addend A number to be added to another number. In $3 + 5 - 8$, the addends are 3, 5, and -8. Also called *summand*.

Algebra A mathematical discipline that generalizes arithmetic with the use of letters to represent numbers.

Algebraic expression A combination of letters and numbers connected by mathematical symbols. For example, $-3x^2 + 4x - 7$, $-q + \sqrt{7r - 1}$, and $\frac{15a + 2}{6b - 8}$ are algebraic expressions.

Alternate exterior angles See *Transversal*.

Alternate interior angles See *Transversal*.

Angle Two line segments with a common endpoint. Either or both segments may be extended endlessly in the direction away from the common endpoint.

Area A number, measured in square units, giving the size of the inside of a plane figure.

Associative property of addition When three or more numbers are added, the grouping of these numbers does not change the resulting sum. In general $(a + b) + c = a + (b + c)$, so that $(6 + 5) + 9 = 6 + (5 + 9)$.

Associative property of multiplication When three or more numbers are multiplied, the grouping of those numbers does not change the resulting product. In general, $(a \times b) \times c = a \times (b \times c)$, so that $(2 \times 3) \times 7 = 2 \times (3 \times 7)$.

Asymptote A curve that extends infinitely in a given direction in such a way that it becomes close to a line is said to have that line as its asymptote in that given direction.

Average The number obtained by dividing the sum of two or more addends by the number of addends. Thus, the average of 2, 8, and -1 is $\frac{2 + 8 - 1}{3} = 3$. Also called *mean*.

Axes See *Cartesian coordinate system*.

Axis of symmetry (of a parabola) The line through which a parabola is reflected; that is, the line of symmetry for the parabola.

Base The number a in a^b. In 2^3, the number 2 is the base. $2^3 = 2 \times 2 \times 2$. The base is used 3 times as a factor.

Binomial A mathematical expression that consists of two terms connected by either addition or subtraction. $3x + 2$, $5x^2 - 20$, and $7y + 19$ are examples of binomials.

Cartesian coordinate system Two perpendicular number lines, called *axes*, used to locate points in a plane. The axes intersect at zero (the *origin*), with positive numbers going up on the vertical line, called the *y-axis*, and to the right on the horizontal line, or *x-axis*. Together, the axes divide the plane into four parts called *quadrants*. Also called *rectangular coordinate system*.

391

Celsius A temperature scale in which 0° represents the freezing point and 100° represents the boiling point of water.

Center See *Circle*; *Sphere*.

Circle A plane figure with all of its points the same distance from a fixed point called its *center*. See *Sphere*.

Circumference The distance around a circle. The circumference of a circle can be found by multiplying its diameter times π ($\pi \approx 3.14$ or $\frac{22}{7}$).

Coefficient Used in this book to represent the numerical coefficient. That is, the coefficient is a number or a letter representing a fixed number in a term. The coefficients of the terms $3x$, $-\frac{2}{5}xy$, and ax^2 are 3, $-\frac{2}{5}$, and a, respectively.

Common denominator A common multiple of two or more denominators. A common denominator for $\frac{1}{3}$ and $\frac{3}{7}$ is 21; another is 42.

Common divisor See *Common factor*.

Common factor A number that is a factor of two or more numbers. The common factors of 8 and 12 are 1, 2, and 4. Also called *common divisor*.

Common logarithm A logarithm with base ten, written $y = \log x$.

Common multiple A number that is a multiple of two or more numbers. A common multiple of 3 and 9 is 27; another is 18.

Commutative property of addition The order in which numbers are added does not change the sum. In general, $a + b = b + a$ so that $2 + 3 = 3 + 2$.

Commutative property of multiplication The order in which numbers are multiplied does not change the product. In general, $a \times b = b \times a$, so that $3 \times 2 = 2 \times 3$.

Complementary angles Two angles are complementary if the sum of their measures is 90°.

Complex fraction A fraction in which either the numerator or denominator (or both) contains a fraction. Also called *compound fraction*.

Complex numbers The square root of negative one ($\sqrt{-1}$) is defined to be the *imaginary unit* (i) and by use of this unit, the complex numbers are constructed to be of the form $a + bi$ where a and b are real. Also called *imaginary numbers*.

Composite number A whole number, greater than 0, that has more than two divisors. For example, 6 is a composite number since its divisors are 1, 2, 3, and 6. The numbers $4, 6, 8, 9, 10, \ldots$ are composite numbers. Compare *Prime number*.

Compound fraction See *Complex fraction*.

Congruent Two figures are congruent if they have the same shape and size. Angles are congruent if they have the same measure.

Conic section The result of the intersection of a plane and a right circular cone is called a conic section. The circle, ellipse, parabola, and hyperbola are common conic sections.

Consistent system of equations A system of equations that has at least one solution.

Constant A number or a letter that represents a fixed number, as distinguished from a variable, which may represent more than one number.

Constant function A function of the form $f(x) = b$, where b is a fixed number.

Corresponding angles See *Transversal*.

Counting numbers See *Natural numbers*.

Cross-product For the fractions $\frac{2}{3}$ and $\frac{5}{8}$, the cross-products are 2×8 and 3×5. In general, the cross-products of $\frac{a}{b}$ and $\frac{c}{d}$ are ad and bc.

Cube A solid object with six equal square faces:

Cylinder A solid figure with circular bases and parallel sides:

Decimal system A system of numeration and computation that is based on the number ten.

Degree The degree of a monomial is the sum of the exponents of the variable(s). The degree of a polynomial is the degree of the highest-degree monomial. The 0 monomial is not assigned a degree.

Denominator The bottom number in a fraction that gives the number of parts into which the numerator is divided. In $\frac{5}{6}$, the number 6 is the denominator.

Dependent system of equations A system of equations in which one equation is redundant, in the sense that the system can be written in fewer equations.

Dependent variable In an equation, when one variable, say y, is expressed in terms of one or more other variables, then y is the dependent variable. In the equations $y = mx + b$, $A = \pi r^2$, and $I = Prt$, the dependent variables are y, A, and I, respectively.

Diagonal A straight line segment that connects one corner of a polygon to another corner, but is not a side of the polygon.

Diameter The maximum distance across a circle; also, the line segment that passes through the center of a circle and whose endpoints lie on the circle. The diameter is equivalent to two radii.

Difference The difference is the result of subtraction. In $10 - 2 = 8$, the number 8 is the difference.

Digit Any whole number from 0 to 9, inclusive.

Dimensional analysis The use of units as if they were themselves numbers. For example,
$$\frac{12 \text{ in.}}{\text{ft.}} \times \frac{3 \text{ ft.}}{\text{yd.}} = \frac{12 \times 3 \text{ in.}}{\text{yd.}} = \frac{36 \text{ in.}}{\text{yd.}}.$$

Discriminant See *Quadratic formula*.

Distributive property In general, $a \times (b + c) = (a \times b) + (a \times c)$, so that $2 \times (5 + 3) = (2 \times 5) + (2 \times 3)$.

Dividend A number that is divided by another number. For example, in $20 \div 4 = 5$, the number 20 is the dividend.

Divisor A number that evenly divides another number. For example, in $20 \div 4 = 5$, the number

4 is the divisor. The divisors of 12 are 1, 2, 3, 4, 6, and 12. See *Factor*.

Domain The set of values that sensibly replaces the independent variable in an equation or function.

e See *Natural base*.

Ellipse A conic section that consists of all points in a plane such that for each point on the curve, the sum of its distances from two fixed points, called the *foci*, is constant. The standard form of an ellipse, centered at the origin, is $\frac{x^2}{a^2} + \frac{y^2}{b^2} = 1$, where $(a, 0)$, $(-a, 0)$, $(0, b)$, and $(0, -b)$ are its *vertices*.

Equation A mathematical sentence using the $=$ symbol.

Equilateral triangle A triangle with all sides equal.

Even number An integer that has a divisor of 2, such as 2, -2, 4, -8, and also 0. Any whole number plus itself results in an even number.

Expanded form The representation of a number as a polynomial in the base 10 with digital coefficients. Thus, the expanded form of 2561 is $2 \times 10^3 + 5 \times 10^2 + 6 \times 10 + 1$.

Exponent The number b in a^b. In 2^3, the number 3 is the exponent. $2^3 = 2 \times 2 \times 2$. A positive exponent integer tells how many times the base (2 in this example) is used as a factor.

Exponential function A function of the form $f(x) = a^x$, where a is a positive constant different from 1.

Factor (1) Any two or more numbers that form a product when multiplied together. (2) A number that evenly divides a second number is a factor of that number. (3) A divisor.

Fahrenheit A temperature scale in which 32° represents the freezing point and 212° the boiling point of water.

Foci See *Ellipse*; *Hyperbola*.

FOIL Acronym for multiplying two binomials: *F*irst *O*uter *I*nner *L*ast, referring to the order in which terms are multiplied.

Fraction A number that can be written in the form $\frac{a}{b}$, such as $\frac{1}{2}$, $\frac{2}{3}$, or $-\frac{7}{5}$.

Function A rule f such that: given a number x, there arises a uniquely-determined number $f(x)$. Every function can be represented by a set of ordered pairs where the first component is associated with exactly one second component.

Geometry A branch of mathematics that deals with the relationships, properties, and measurement of points, lines, angles, surfaces, and solids.

Graph A pictorial representation of a function, usually in the Cartesian coordinate system, where each ordered pair of the graph satisfies the function.

Hexagon A polygon with six sides.

Hyperbola A conic section that consists of all points in a plane such that for each point on the curve, the difference of the distances from two fixed points, called the *foci*, is constant. The standard form of a hyperbola, centered at the origin, is $\frac{x^2}{a^2} - \frac{y^2}{b^2} = 1$ (or $\frac{y^2}{b^2} - \frac{x^2}{a^2} = 1$), where $(a, 0)$ and $(-a, 0)$ (or $(0, b)$ and $(0, -b)$) are its *vertices*.

Hypotenuse The longest side of a right triangle, always located opposite the right angle.

i See *Complex numbers*.

Identity In addition, the identity is the number zero, which leaves addition unchanged when used as an addend. That is, $n + 0 = n$ for all numbers n. In multiplication, the identify is the number 1, which leaves multiplication unchanged when used as one of the factors. That is, $n \times 1 = n$ for all numbers n.

Imaginary numbers See *Complex numbers*.

Imaginary unit See *Complex numbers*.

Improper fraction A fraction whose numerator is greater than its denominator.

Inconsistent system of equations A system that has no solution.

Independent system of equations A system that has a unique solution, such as two lines with different slopes, *or* no solution, such as a pair of parallel lines.

Independent variable In an equation, if one variable is solved for in terms of another variable, for example x, then x is the independent variable. In the equations $y = ax^2 + bx + c$, $A = \pi r^2$, and $C = \pi d$, the independent variables are x, r, and d, respectively.

Index (of a radical) In the expression $\sqrt[n]{a}$, (the nth root of a), n is called the index of the radical.

Inequality A statement of order between mathematical expressions, involving the symbols $<$, $>$, $\leq$, or $\geq$.

Integers The positive and negative whole numbers, along with 0.

Inverse In addition, the opposite: the inverse of 2 is -2; the inverse of -5 is 5; the inverse of 0 is 0. In multiplication, the reciprocal: the inverse of 2 is $\frac{1}{2}$; the inverse of $\frac{2}{3}$ is $\frac{3}{2}$; 0 has no inverse. Otherwise, inverse is generally taken to be the reverse: the inverse of taking a square root is squaring, for example.

Irrational number A number that cannot be expressed as an integer or as a quotient of two integers. In decimal form, an irrational number is nonrepeating and nonterminating.

Isosceles triangle A triangle with two equal sides.

Least common denominator (LCD) The least (positive) common multiple of the denominators of two or more fractions.

Least common multiple (LCM) The least common multiple of two or more numbers is the smallest positive number that has each as a factor. The LCM of 4, 6, and 8 is 24.

Like terms Terms that differ only by their coefficients. Also called *similar terms*.

Line A straight geometric plane figure that connects two distinct points and extends infinitely in both directions.

Linear equation A first-degree equation.

Linear function A function whose graph is a straight line.

Line segment The part of a line that connects and lies between two points.

Literal coefficient A coefficient that is represented by a letter which stands for a fixed number.

Logarithm function For $a > 0$, $a \neq 1$, $x = a^y$ is equivalent to $y = \log_a x$. The logarithm function is the inverse of the exponential function.

Mean See *Average*.

Minuend The number or quantity from which another number is to be subtracted. In $10 - 2 = 8$, the number 10 is the minuend.

Mixed number A number that has a whole number part and a fraction part, such as $4\frac{2}{5}$.

Monomial A constant, a variable, or a product of constants and variables. A polynomial with one and only one term, such as $4x$, $-3y^3$, or $\frac{2}{5}x^2y$.

Multiple A number that is a product of some specified number and another number. Some multiples of 5 are 5, 10, 15.

Natural base A constant, used in the natural logarithm, that is approximately 2.7182818 and is defined by $\lim_{n \to \infty} (1 + \frac{1}{n})^n = e$.

Natural logarithm A logarithm with base e, written $y = \ln x$.

Natural numbers The set N of positive integers: $N = \{1, 2, 3, \ldots\}$. Also called the *counting numbers*.

Negative number A number is negative if it lies to the left of zero on a number line. For example, -7, -3.4, $-\frac{1}{2}$, and -2.046 are negative numbers.

Number line A visual representation of the positions of numbers:

Numerator The top number in a fraction that gives the amount to be divided. In the fraction $\frac{7}{9}$, the number 7 is the numerator.

Octagon A polygon with eight sides.

Odd number An integer that does not have a factor of 2, such as -5, 3, 7, or -21. Any whole number plus itself plus 1 results in an odd number.

Opposite numbers Two numbers whose sum is zero. The numbers $+7$ and -7 are opposites because $(+7) + (-7) = 0$.

Ordered pair A pair of numbers representing a point in the Cartesian coordinate system. An ordered pair is written in the form (x, y).

Ordinate The second element of an ordered pair, (x, y), denoting the vertical coordinate. See *Abscissa*.

Origin The point that designates the number zero on a number line. In two dimensions, it is the point of intersection of the coordinate axes.

Parabola The set of points in a plane that satisfy the quadratic equation $y = ax^2 + bx + c$, or $x = ay^2 + by + c$, $a \neq 0$.

Parallel lines Lines in the same plane that do not meet; that is, do not have a point in common.

Parallelogram A quadrilateral with opposite sides parallel and of equal length.

Percent A word indicating "hundredths" or "out of 100." For example, 52 percent (52%) means $\frac{52}{100}$ or 0.52.

Perimeter The sum of the lengths of the sides of a polygon.

Pi (π) The number obtained by dividing the circumference of any circle by its diameter. Common approximations for pi are 3.14, $\frac{22}{7}$, and $\frac{355}{113}$.

Plane A flat surface.

Plane figure A two-dimensional representation of points lying in a plane, such as a rectangle or circle.

Polygon A closed plane figure with three or more sides that are line segments.

Polynomial An algebraic expression that can be written as a monomial or a sum of monomials.

Positive numbers A number is positive if it lies to the right of zero on a number line. For example, 3, 21.7, $\frac{4}{9}$, and $3\frac{1}{2}$ are positive numbers.

Power A meaningful expression of the form a^b. For example, 5^3 is a power, sometimes called the third power of 5. $5^3 = 5 \times 5 \times 5$. Here, 5 is used 3 times as a factor. 5 is the base and 3 is the exponent.

Prime number A prime number is a whole number greater than 1 whose only factors are itself and

1. The numbers $2, 3, 5, 7, 11, \ldots$ are prime numbers. Compare *composite number*.

Prism A solid with two faces that are polygons in parallel planes and the other faces parallelograms.

Product A product is the result of multiplying two or more numbers. In $7 \times 5 = 35$, the number 35 is the product.

Proper fraction A fraction whose numerator is less than its denominator.

Proportion A statement that two ratios are equal, such as $\frac{2}{3} = \frac{8}{12}$. In general, for real numbers a, b, c, and d, $\frac{a}{b} = \frac{c}{d}$ is a proportion (b and d nonzero).

Pyramid A solid whose base is a polygon and whose faces are triangles with a common vertex.

Pythagorean theorem In a right triangle, where a and b are the lengths of the legs of the triangle and c is the length of the hypotenuse, then $a^2 + b^2 = c^2$.

Quadrant See *Cartesian coordinate system*.

Quadratic equation For real numbers a, b, and c, $ax^2 + bx + c = 0$, $a \neq 0$, is called a quadratic equation.

Quadratic formula Used to find roots (or solutions) to quadratic equations, $ax^2 + bx + c = 0$: $x = \dfrac{-b \pm \sqrt{b^2 - 4ac}}{2a}$. The radicand, $b^2 - 4ac$, is called the *discriminant* of the quadratic.

Quadrilateral A four-sided polygon.

Quotient The quotient is the result of division. In $63 \div 7 = 9$, the number 9 is the quotient.

Radicand An algebraic expression to be evaluated inside the square root symbol ($\sqrt{}$). In $\sqrt{b}$, b is the radicand.

Radius The distance from the center of a circle to its edge. The radius of a circle is one half the diameter. Plural is *radii*.

Range The set of values that meaningfully replace the dependent variable in a function.

Ratio A comparison of two numbers or quantities by division. For example, the ratio 3 to 5 can be written $\frac{3}{5}$ or $3:5$.

Rational number A number that can be expressed as an integer or as a quotient of two integers. In decimal form, a rational number is repeating or terminating.

Real numbers The set of all rational and irrational numbers.

Reciprocals Two numbers whose product is 1. The numbers $\frac{2}{3}$ and $\frac{3}{2}$ are reciprocals because $\frac{2}{3} \times \frac{3}{2} = 1$. Likewise, $\frac{2}{3}$ is called the reciprocal of $\frac{3}{2}$.

Rectangle A parallelogram with four right angles.

Rectangular coordinate system See *Cartesian coordinate system*.

Rectangular prism A solid object with six rectangular faces, like a box.

Remainder The amount that is left over when one number does not evenly divide another. In $7\overline{)15}$ $\frac{14}{1}$, the number 1 is the remainder.

Right angle An angle that has a measure of $90°$, like the corner of a book.

Right triangle A triangle with a right angle.

Root of a number The expression $\sqrt{a}$, square root of a, means a number b, such that $b \times b = b^2 = a$. The expression $\sqrt[3]{a}$, cube root of a, means a number c such that $c \times c \times c = c^3 = a$. In general, $\sqrt[n]{a}$ (nth root of a) means a number x such that $x^n = a$.

Root of an equation See *Solution*.

Scientific notation Writing a number as the product of a number between 1 and 10 and a power of 10. For example, the number 2,300,000 can be written in scientific notation as 2.3×10^6.

Set A collection of objects.

Similarity Two geometric figures are said to be similar if they have the same shape but are not necessarily the same size.

Similar terms See *Like terms*.

Slope of a line A numerical representation of the steepness of a line. The slope of a line containing the points (x_1, y_1) and (x_2, y_2), denoted by m, is defined as

$$m = \frac{y_2 - y_1}{x_2 - x_1} = \frac{\text{Vertical change}}{\text{Horizontal change}} = \frac{\text{Rise}}{\text{Run}}$$

Solution The value(s) or ordered pair(s) that make an equation, inequality, or system of equations true; also called *root*.

Sphere A space figure with all of its points equally distant from a given fixed point called its center. See *Circle*.

Square A rectangle with all sides equal.

Square number A number that is the product of a number and itself. 49 is a square number since $49 = 7 \times 7$, written 7^2. The numbers $0, 1, 4, 9, 16, 25, 36, \ldots$ are square numbers.

Square root x is a square root of n if $x \times x = n$. For example, 8 is a square root of 64 since $8 \times 8 = 64$; -8 is also a square root of 64 because $(-8)(-8) = 64$.

Subtrahend A number or quantity to be subtracted from another. In $10 - 2 = 8$, the number 2 is the subtrahend.

Sum A sum is the result of adding two or more numbers. In $12 + 5 = 17$, the number 17 is the sum.

Summand See *Addend*.

Supplementary angles Two angles are supplementary if the sum of their measures is 180°.

System of equations A set of two or more equations that are to be solved simultaneously.

Terms In $4x - 3y + 7$, the terms are $4x$, $-3y$, and 7. See *Addend*.

Transversal A line that intersects a system of lines. In the diagram, lines l and m are cut by a

transversal t:

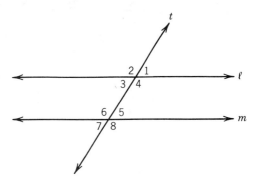

In doing so, eight angles are formed and can be paired in different categories:
  alternate exterior angles: 1 and 7, 2 and 8.
  alternate interior angles: 3 and 5, 4 and 6.
  corresponding angles: 1 and 5, 2 and 6, 3 and 7, 4 and 8.
  vertical angles: 1 and 3, 2 and 4, 5 and 7, 6 and 8.

Triangle A three-sided polygon.

Trinomial A polynomial containing exactly three terms.

Variable A letter that represents one or more of several possible numbers in a mathematical equation or expression.

Vertex of a parabola The point where the axis of symmetry crosses a parabola.

Vertical angles See *Transversal*.

Vertices See *Ellipse*; *Hyperbola*. Singular is *vertex*.

Volume A number, measured in cubic units, giving the size of the inside of a solid object.

Whole number The numbers $0, 1, 2, 3, \ldots$ are whole numbers.

x-axis See *Cartesian coordinate system*.

x-intercept The point(s) where a graph passes through the x-axis, always of the form $(x, 0)$.

y-axis See *Cartesian coordinate system*.

y-intercept The point(s) where a graph passes through the y-axis, always of the form $(0, y)$.

ANSWERS TO EVEN-NUMBERED PROBLEMS

A1.1, PAGES 5–6

2. 5, 8, 3   **4.** 4, 3, 7   **6.** 7, 3, 9, 8   **8.** 1, 1, 1, 1   **10.** 8, 3, 0, 2
12. 3, 0, 0, 4   **14.** 1, 2, 3, 4   **16.** 532   **18.** 1030   **20.** 3506
22. 306,306,306; three hundred six million three hundred six thousand three hundred six
24. $(8 \times 100) + (9 \times 10)$ **26.** $(3 \times 1000) + (2 \times 1)$
28. $(9 \times 10,000) + (9 \times 1000) + (3 \times 100)$ **30.** $(3 \times 100,000)$
32. $(5 \times 10,000,000) + (8 \times 1,000,000) + (5 \times 1)$

A1.2, PAGES 9–10

2. 770   **4.** 1213   **6.** 1618   **8.** 831   **10.** 625   **12.** 910
14. 1023   **16.** 741   **18.** 1405   **20.** 1369   **22.** 1951   **24.** 15,487
26. 20,000   **28.** 62   **30.** 80   **32.** 179   **34.** 186   **36.** 489
38. 2847   **40.** 1888   **42.** 3087   **44.** 6174   **46.** 1848   **48.** 6389
50. 822   **52.** 939   **54.** 280

A1.3, PAGES 13–14

2. 850   **4.** 13,482   **6.** 58,291   **8.** 32,572   **10.** 8289   **12.** 58,305
14. 12,780   **16.** 37,884   **18.** 22,360   **20.** 265,174   **22.** 246,140
24. 822,402   **26.** 353,421   **28.** 13   **30.** 21   **32.** 26   **34.** 111
36. 395   **38.** $202r24$ **40.** $55r9$ **42.** $705r5$ **44.** $84r203$
46. $384r41$ **48.** $34r10$

A1.4, PAGES 17–18

2. $+10$ **4.** $+5$ **6.** -6 **8.** $+27$ **10.** -50 **12.** $+200$
14. -49 **16.** -60 **18.** $+1100$ **20.** -41 **22.** $+1$ **24.** -9

26. $+16$ **28.** -14 **30.** $+69$ **32.** $+106$ **34.** -1 **36.** -300
38. -296 **40.** $+45$ **42.** $+10$ **44.** $+72$ **46.** -24
48. -24 **50.** $+25$ **52.** $+28$ **54.** -44 **56.** -1000 **58.** 0
60. $+160$ **62.** $+3$ **64.** $+2$ **66.** $+2$ **68.** -4 **70.** $+2$
72. $+3$ **74.** -5 **76.** -3 **78.** $+4$ **80.** $+4$

A1.5, PAGES 21 – 22

2. 7 **4.** -25 **6.** 1 **8.** 0 **10.** 64 **12.** 3 **14.** -3 **16.** 8
18. -13 **20.** 4 **22.** 6 **24.** 25 **26.** -76 **28.** 16 **30.** 16
32. 0 **34.** 99 **36.** 17 **38.** 162 **40.** 32 **42.** 16 **44.** 0
46. 10 **48.** 14 **50.** 1 **52.** 47 **54.** 0 **56.** 9 **58.** -21
60. 47

A1.6, PAGES 25 – 26

2. $>$ **4.** $>$ **6.** $=$ **8.** $>$ **10.** $>$ **12.** is greater than
14. □ is less than □□   **16.** $\ldots, -1, 0, 1, 2, 3, 4$ **18.** $8, 9, 10, \ldots$
20. $\ldots, -1, 0, 1, 2, 3$ **22.** $3, 4, 5$ **24.** $-4, -3, -2, \ldots, 7, 8, 9$
26. $201, 202, 203, \ldots, 504, 505, 506$ **28.** $2, 3, 5, 7, 9$
30. $29, 34, 89, 101, 102$ **32.** $70, 67, 50, 42, 39, 31$
34. Answers will vary   **36.** $0, 1, 2, 3, 4$ **38.** $0, 1, 2, 3, 4$
40. There is none

A2.1, PAGES 33 – 34

2. 20 **4.** 21 **6.** 32 **8.** 27 **10.** 34 **12.** 52 **14.** $\dfrac{19}{30}$

16. $3\dfrac{9}{20}$ **18.** $1\dfrac{3}{4}$ **20.** $\dfrac{1}{2}$ **22.** $\dfrac{2}{3}$ **24.** $\dfrac{11}{7}$ **26.** $2\dfrac{2}{3}$

28. $2\dfrac{3}{10}$ **30.** $1\dfrac{14}{18} = 1\dfrac{7}{9}$ **32.** 3 **34.** $2\dfrac{26}{28} = 2\dfrac{13}{14}$ **36.** $1\dfrac{27}{36} = 1\dfrac{3}{4}$

38. $3\dfrac{13}{26} = \dfrac{1}{2}$ **40.** $4\dfrac{1}{55}$ **42.** $9\dfrac{62}{93} = 9\dfrac{2}{3}$ **44.** $\dfrac{22}{9}$ **46.** $\dfrac{40}{9}$

48. $\dfrac{73}{11}$ **50.** $\dfrac{26}{3}$ **52.** $\dfrac{87}{8}$ **54.** $\dfrac{90}{7}$ **56.** $\dfrac{117}{8}$ **58.** $\dfrac{167}{10}$

60. $\dfrac{245}{13}$ **62.** $\dfrac{247}{12}$ **64.** $\dfrac{388}{17}$ **66.** $\dfrac{397}{16}$ **68.** $\dfrac{350}{13}$

70. $\dfrac{859}{30}$ **72.** $\dfrac{803}{26}$

A2.2, PAGES 37 – 38

2. $\dfrac{3}{16}$ **4.** $\dfrac{8}{15}$ **6.** $\dfrac{8}{15}$ **8.** $\dfrac{1}{4}$ **10.** $\dfrac{2}{5}$ **12.** $\dfrac{1}{8}$ **14.** $\dfrac{2}{5}$

16. $\dfrac{3}{10}$ **18.** $\dfrac{7}{16}$ **20.** $-\dfrac{5}{8}$ **22.** $\dfrac{1}{12}$ **24.** $\dfrac{1}{2}$ **26.** $3\dfrac{1}{2}$

28. 24   **30.** $3\frac{5}{6}$ **32.** 4   **34.** 28   **36.** 12   **38.** 10   **40.** 15

42. $10\frac{1}{2}$ **44.** $10\frac{4}{5}$ **46.** $\frac{8}{9}$ **48.** $2\frac{1}{10}$ **50.** $1\frac{2}{3}$ **52.** $-\frac{2}{7}$

54. $-\frac{2}{5}$ **56.** $2\frac{1}{5}$ **58.** $9\frac{1}{3}$ **60.** $4\frac{1}{2}$ **62.** $2\frac{4}{7}$ **64.** $\frac{5}{7}$

66. $\frac{1}{3}$ **68.** $5\frac{1}{4}$ **70.** $-\frac{2}{9}$ **72.** $\frac{1}{18}$ **74.** $4\frac{6}{7}$ **76.** 4

78. $1\frac{4}{7}$ **80.** $\frac{2}{13}$ **82.** $-1\frac{26}{75}$ **84.** $-\frac{7}{34}$ **86.** 2   **88.** $\frac{1}{6}$

90. -24

A2.3, PAGES 41–42

2. $\frac{5}{7}$ **4.** $\frac{1}{6}$ **6.** $\frac{2}{3}$ **8.** $-\frac{5}{8}$ **10.** $4\frac{1}{4}$ **12.** $-1\frac{2}{7}$

14. $-1\frac{1}{3}$ **16.** $-\frac{1}{3}$ **18.** $2\frac{2}{7}$ **20.** $\frac{8}{9}$ **22.** $\frac{1}{3}$ **24.** $4\frac{5}{9}$

26. $2\frac{1}{8}$ **28.** $-2\frac{1}{22}$ **30.** $\frac{11}{36}$ **32.** $-\frac{31}{42}$ **34.** $-\frac{7}{30}$

36. $1\frac{1}{12}$ **38.** $\frac{17}{30}$ **40.** $-1\frac{11}{24}$ **42.** $2\frac{17}{30}$ **44.** $\frac{37}{60}$

46. $-\frac{29}{63}$ **48.** $-\frac{21}{50}$ **50.** $2\frac{1}{2}$ **52.** $8\frac{1}{15}$ **54.** $-\frac{7}{9}$ **56.** $2\frac{1}{36}$

58. $3\frac{17}{42}$ **60.** 7   **62.** $5\frac{7}{10}$ **64.** $-6\frac{1}{24}$ **66.** $\frac{121}{126}$

68. $10\frac{79}{100}$ **70.** $7\frac{142}{315}$

A3.1, PAGES 45–46

2. 0.007   **4.** 0.12   **6.** 1,000,715   **8.** 0.0025   **10.** 5001.012
12. hundreds   **14.** ones   **16.** tens   **18.** hundredths
20. millionths   **22.** 0.03   **24.** 0.0037   **26.** 0.264   **28.** 2.007
30. 0.00054   **32.** 0.6   **34.** 0.9   **36.** 15.9   **38.** 106.8
40. 0.08   **42.** 0.95   **44.** 4.36   **46.** 0.63   **48.** 0.111
50. 0.779   **52.** 17.626   **54.** 5.253   **56.** 37.100

A3.2, PAGES 49–50

2. 6.7   **4.** 8.885   **6.** 14.61   **8.** 4.29   **10.** 11.405   **12.** 0.555
14. 0.965   **16.** 217.13   **18.** 76.22   **20.** $321.00   **22.** 511.405
24. 79.197   **26.** 1.14   **28.** 4.56   **30.** 53.19   **32.** $26.66
34. $228.56   **36.** $24.79   **38.** 114.107   **40.** 0.364   **42.** $5.39
44. 0.69   **46.** 0.23   **48.** 3.995

A3.3, PAGES 55–56

2. 52.6 **4.** 1.141 **6.** 0.625 **8.** 0.25 **10.** 7.14 **12.** 0.0729
14. 10.608 **16.** 0.15846 **18.** 0.001716 **20.** 10.4176 **22.** 5.27
24. 21.2 **26.** $0.23 **28.** 0.091 **30.** 0.609 **32.** 0.64 **34.** 2.1
36. 3.1 **38.** 148 **40.** 1.5 **42.** 700 **44.** 203 **46.** 1,120
48. 5.15 **50.** 33.11 **52.** 0.882 **54.** 0.765 **56.** 14.8

A4.1, PAGES 63–64

2. 16 **4.** 343 **6.** 81 **8.** 64 **10.** 10,000 **12.** 100,000

14. 6561 **16.** 243 **18.** $\dfrac{256}{2401}$ **20.** $\dfrac{6561}{256}$ **22.** 0.027

24. 15.625 **26.** $\dfrac{7}{12}$ **28.** $\dfrac{13}{81}$ **30.** 5^6 or 15,625 **32.** 2^9 or 512

34. 4^8 or 65,536 **36.** 9 **38.** 100 **40.** 81 **42.** 125 **44.** 216

46. 702 **48.** 41 **50.** 82 **52.** $\dfrac{41}{625}$

A4.2, PAGES 67–68

2. 5 **4.** 7 **6.** 11 **8.** 15 **10.** 19 **12.** 22 **14.** 16 **16.** 1

18. 0.4 **20.** 0.03 **22.** $\dfrac{3}{7}$ **24.** $\dfrac{4}{5}$ **26.** 2 **28.** 3, 4

30. 7, 8 **32.** 9, 10 **34.** 15, 16 **36.** 24, 25 **38.** 20, 21
40. 8, 9 **42.** 89, 90 **44.** 7, 8 **46.** 70, 71 **48.** 31, 32
50. 3162, 3163

A5.1, PAGES 73–74

2. 3.4 **4.** 6.75 **6.** 1.55 **8.** 2.5 **10.** 2.84 **12.** 0.66 **14.** 2.6
16. 2.25 **18.** 1.36 **20.** 0.232 **22.** 0.175 **24.** 0.012
26. 5.375 **28.** 3.25 **30.** 5.6875 **32.** 7.45 **34.** 2.68
36. 0.328 **38.** 0.075 **40.** 3.125 **42.** 0.0032 **44.** 0.15625
46. 0.016 **48.** 0.00032 **50.** 0.65 **52.** 0.3 **54.** 0.015
56. 1.7 **58.** 5.007 **60.** 8.9 **62.** 2.099 **64.** 0.379 **66.** 3.58
68. 1.5 **70.** 4.9

A5.2, PAGES 77–78

2. $\dfrac{1}{2}$ **4.** $\dfrac{4}{25}$ **6.** $\dfrac{5}{4}$ **8.** $\dfrac{126}{125}$ **10.** $\dfrac{25}{4}$ **12.** $\dfrac{64}{625}$ **14.** $\dfrac{5}{8}$

16. $\dfrac{1}{8}$ **18.** $\dfrac{32}{25}$ **20.** $\dfrac{5}{32}$ **22.** $\dfrac{130,213}{2000}$ **24.** $\dfrac{120,421}{2500}$

26. $\dfrac{23}{2500}$ **28.** $\dfrac{15}{32}$ **30.** $\dfrac{19}{160}$ **32.** $\dfrac{1957}{80}$ **34.** $\dfrac{345,679}{1000}$

36. $\dfrac{654,321}{1000}$ **38.** $\dfrac{527}{6250}$ **40.** $\dfrac{1}{128}$ **42.** $\dfrac{1}{40,000}$

44. $\dfrac{4,000,002}{78,125}$ **46.** $1\dfrac{7}{25}$ **48.** $1\dfrac{1}{200}$ **50.** $2\dfrac{14}{25}$ **52.** $1\dfrac{3}{25}$

54. $2\dfrac{161}{400}$ **56.** $437\dfrac{5}{16}$ **58.** $64\dfrac{204}{625}$ **60.** $85\dfrac{717}{2000}$ **62.** 5

64. $\dfrac{175}{16}$, 10.9375

A6.1, PAGES 81 – 82

2. 23 circulars **4.** 204 lbs. **6.** 2.5 goals **8.** 26° **10.** 11.5 books

A6.2, PAGES 85 – 88

2. 552% **4.** 0.0552 **6.** $\dfrac{69}{1250}$ **8.** 75% **10.** 0.0075 **12.** $\dfrac{3}{400}$
14. 1% **16.** 70% **18.** 37.5% **20.** 6% **22.** 301 students
24. 40% **26.** $12.60 **28.** 120% **30.** $171,540 **32.** 60%
34. $32.79

A6.3, PAGES 91 – 94

2. $3.01 **4.** $4.47 **6.** $12\dfrac{1}{2}$ lbs. **8.** 52,235 gallons **10.** 1868
12. 4 : 45 p.m. **14.** $187.20

B1.1, PAGES 103 – 104

2. $-8 + n$ or $n + (-8)$ **4.** $9k$ **6.** $5pq$ **8.** $6m(-n)$ or $-6mn$
10. $-8s + (-3t)$ **12.** yes, yes **14.** no, yes **16.** yes, yes
18. no, yes **20.** $7t$ **22.** $16s$ **24.** $5x$ **26.** $-m$ **28.** $2r + 3s$
30. $-4p - q$ **32.** $5b - 5n$ **34.** $13x + 2y$
36. $-6b - 3m - 2n$ **38.** $4k - 5ab$ or $-5ab + 4k$ **40.** $5x + 2y$
42. $-3p + pq$ or $pq - 3p$ **44.** $4s - t$ **46.** $\dfrac{1}{3}x - \dfrac{2}{3}y$
48. $\dfrac{11}{6}m - \dfrac{13}{12}n$ **50.** $2x - 4xy + 3xyz$

B1.2, PAGES 107 – 108

2. 10 **4.** 3 **6.** 4 **8.** 4 **10.** 21 **12.** 16 **14.** 0 **16.** -19
18. 55 **20.** 7 **22.** -68 **24.** -21 **26.** 33 **28.** -14
30. 60 **32.** -44 **34.** 7 **36.** 14 **38.** 25 **40.** 32 **42.** 184
44. 168 **46.** 29 **48.** 29

B1.3, PAGES 111 – 112

2. $6x^2 + 6x + 9$ **4.** $9x^2 + 2x + 4$ **6.** $4y^2 + y + 4$
8. $4u^2 - 2u - 2$ **10.** $4v^2 + v$ **12.** $7w^2 + 2$ **14.** t **16.** 0
18. $2x^2 + 4x - 8$ **20.** $4x^2 + 2x + 8$ **22.** $t^2 - 3t - 1$
24. $2r^2 - 2r$ **26.** $4a^2 - 2a - 2$ **28.** $-7m^2 + 10m + 15$
30. $-8x^2 + 8x - 12$ **32.** $-t^2 + 8t + 3$ **34.** $5k^3 + 5k^2 - 3k + 3$
36. $-3u^3 + 7u^2 - 14u + 5$ **38.** $-7v^3 - 5v^2 - 3v - 5$
40. $-14x + 18$

B1.4, PAGES 115 – 116

2. a^6 **4.** c^6 **6.** $-9m^4n$ **8.** $15s^4t$ **10.** $-24p^3q^5$ **12.** -432
14. 2430 **16.** 960 **18.** $2x^5 + 5x^4$ **20.** $6r^2 - 8r$
22. $16k^3 - 12k^2$ **24.** $7b^3c + 14bc^3$ **26.** $15t^4 - 20t^5$
28. $18b^3c^2 + 12b^2c^3 - 24b^2c$ **30.** $32t^8 - 24t^6 - 40t^5$
32. $18x^5y^4 - 9x^2y^5 - 54x^3y^6$ **34.** $35p^3q^3r^4 + 49p^2q^5r^5 + 28pq^2r^5$
36. $-35m^7n^5 - 15m^9n^2 + 25m^5n^2p^3$
38. $-36a^9b^8c^8 + 28a^8b^4c^3 + 32a^6b^7c^3 + 4a^6b^4c^9$
40. $-52g^{14}h^{14}k^3 - 76g^6h^{13}k^2 + 64g^{15}h^9k^7 - 72g^6h^9k^{12}$

B1.5, PAGES 121 – 122

2. $y^2 + 7y + 10$ **4.** $3x^2 + 13x + 12$ **6.** $3m^2 - 10mn - 8n^2$
8. $15x^2 - 34xy + 15y^2$ **10.** $9j^2 - 4k^2$ **12.** $35a^2 + 29ab + 6b^2$
14. $-16u^2 + 40uv - 25v^2$ **16.** $6r^2 - 5rst - 4s^2t^2$
18. $42 - 17c - 15c^2$ **20.** $25s^2 - t^2$ **22.** $16a^2 - 25b^2$
24. $49p^2 - 9q^2$ **26.** $28xy$ **28.** $n^6 - m^4$
30. $pr + 5pq + 3qr + 15q^2$ **32.** $3a^3 - 3ab^2 + 2a^2b - 2b^3$
34. $12j^2k^2 + 17j^2k - 5j^2$ **36.** $25m^6 - 4u^2v^4$
38. $-26p^3q^3 - 16p^5q + 35pq^5$
40. $24x^5y^4 + 16x^2y^5z - 27x^4y^2z^2 - 18xy^3z^3$

B1.6, PAGES 125 – 126

2. $s^2 - 6s + 9$ **4.** $u^2 + 10u + 25$ **6.** $16 - 8t + t^2$
8. $u^2 - 2uv + v^2$ **10.** $16u^2 - 24uv + 9v^2$ **12.** $49p^2q^2 - 42pq + 9$
14. $49m^2 + 14mn^2 + n^4$ **16.** $9r^2 + 12rs + 4s^2$
18. $25a^2 + 60abc + 36b^2c^2$ **20.** $16t^2 + 2t + \dfrac{1}{16}$
22. $9s^2 - 3s + \dfrac{1}{4}$ **24.** $\dfrac{1}{x^2} - 2 + x^2$ **26.** $4a^2b^6 - 20a^3b^3 + 25a^4$
28. $p^2q^6 - 10p^3q^3 + 25p^4$ **30.** $16s^6t^2 + 24s^5t^5 + 9s^4t^8$
32. $81u^6v^4w^2 - 90u^4v^4w^4 + 25u^2v^4w^6$
34. $\dfrac{4r^2s^2}{9} + \dfrac{16rst}{3} + 16t^2$ **36.** $42\dfrac{1}{4}$ **38.** $72\dfrac{1}{4}$ **40.** $4xy$

B1.7, PAGES 129–130

2. $3y - 2$ **4.** $5s + 7$ **6.** $2k^2 - 12k$ **8.** $-69q^5 - 6q^2 + 7$

10. $2w + 7$ **12.** $19x^3 - 29x$ **14.** $-15t^2 + 13$ **16.** $-11v^2 - 9v$

18. $-5m^3 + 2m^2 - 2$ **20.** $14p^2 - p + 2$ **22.** $-16x^4 - 12x^2 + 8x$

24. $-16k^3 + 13k - 8$ **26.** $24u^3 - \dfrac{1}{2}u^2 - 2u$

28. $-6y^3 + \dfrac{3}{4}y^2 - \dfrac{1}{2}y - 2$

B1.8, PAGES 133–134

2. $2x - 1$ **4.** $x + 1$ **6.** $x + 2$ **8.** $3x - 1$ **10.** $2x^2 + 6x + 3$

12. $2x^3 - 3$ **14.** $8x^3 + 4x^2 + 2x + 1$ **16.** $x^2 - 2x + 1$

18. $x^3 + 3x^2 + 3x + 1$ **20.** $x^2 - x + 1$ **22.** $x^4 - x^3 + x^2 - x + 1$

B1.9, PAGES 137–138

2. $4(3m - 2)$ **4.** $3(a + b)$ **6.** $4(x^2 + 1)$ **8.** $17p(2p - 3q)$

10. $3s(9r + 14s)$ **12.** $3ab(2b + 3a - 4)$ **14.** $x^2(5x^2 + x + 4)$

16. $3a(1 + 2a + 3a^2 + 4a^3)$ **18.** $3uv(3u^2 - 2uv + 5v)$

20. $13x^4y^4z^3(-4x^3yz^3 + 7y^2 - 5xz)$ **22.** $(p - q)(s + t)$

24. $(2b + 3)(s - 3)$ **26.** $(5 - x)(m - 6n)$ **28.** $(s + t)(u + v)$

30. $(d + 1)(cd - 1)$ **32.** $(m + 3)(m + 6n)$ **34.** $3(3x + 1)(x - 2)$

36. $(6 - n^2)(3 + m)$ **38.** $(3x - 2y)(6x^2 - 5ay)$

B1.10, PAGES 141–142

2. $(x + 3)(x + 1)$ **4.** $(x + 2)^2$ **6.** N.F. **8.** $(4x + 1)^2$

10. $(x - 5)^2$ **12.** $(x - 7)(x + 5)$ **14.** $(x - 4)(x - 2)$

16. $(5x - 6)^2$ **18.** N.F. **20.** $(2x + 5)(x - 2)$ **22.** N.F.

24. $(2x + 3)(8x - 1)$ **26.** N.F. **28.** $(3x + 1)(x - 3)$

30. $7(x - 1)^2$ **32.** $(2x + 1)(x + 3)$ **34.** $(2x - 7)(x + 3)$

36. N.F. **38.** $(4x + 3)(3x - 5)$ **40.** N.F.

B1.11, PAGES 145–146

2. x, y **4.** $b, 5, 4b$ **6.** $12, 1, 12, 1$ **8.** $x^2, z^2, x^2, 5$

10. $2, x, 3y, x, y$ **12.** $12, 1, 1, 1, 1$ **14.** $5, 1, 1, 1, 1$

16. $(8w - 9z)(8w + 9z)$ **18.** $x(5p - 4r^2)(5p + 4r^2)$

20. $5p(p - 1)(p + 1)$ **22.** $2(3a - 4)(3a + 4)$ **24.** N.F.

26. $x^2y^4(1 - y)(1 + y)$ **28.** $(y - x)(y + x)$

30. $(a + b - x - y)(a + b + x + y)$ **32.** N.F.

34. $(2ab - 3x^4)(2ab + 3x^4)$ **36.** $(ab^2 - p^3q^4)(ab^2 + p^3q^4)$

38. $(x - p - q)(x + p + q)$ **40.** 399 **42.** $39{,}999$ **44.** 884

46. 179 **48.** 1025

B2.1, PAGES 155 – 156

2. $\dfrac{1}{3}$ **4.** $-\dfrac{1}{3z}$ **6.** $-\dfrac{1}{2x}$ **8.** $\dfrac{b}{12}$ **10.** $\dfrac{-2p}{3}$ **12.** $\dfrac{3y}{4z}$

14. $\dfrac{3ux}{4}$ **16.** $\dfrac{1}{10p}$ **18.** $-\dfrac{2p}{9}$ **20.** q **22.** $\dfrac{x-y}{x+y}$ **24.** $\dfrac{m^2}{3}$

26. $\dfrac{3}{q+2}$ **28.** $-w-z$ **30.** $-a-3$

B2.2, PAGES 159 – 160

2. 4   **4.** 19   **6.** 0   **8.** 0   **10.** -62 **12.** $1\dfrac{1}{2}$ **14.** -2

16. -1 **18.** $\dfrac{1}{49}$ **20.** 25

B2.3, PAGES 163 – 164

2. $\dfrac{11w}{x}$ **4.** $\dfrac{7a}{2x}$ **6.** $\dfrac{5+8b^2}{ab}$ **8.** $\dfrac{15b+4}{9b^2+6b}$ **10.** $\dfrac{3xy+5z}{15}$

12. $\dfrac{5x^2-3x}{15z}$ **14.** $\dfrac{5-p}{p}$ **16.** $\dfrac{3}{x-3}$

18. $\dfrac{a^2+2ab+b^2+5a-5b}{a^2-b^2}$ **20.** $-\dfrac{2y}{x-2}$

22. $\dfrac{2a+8b}{(a-b)^2(a+b)}$ **24.** $\dfrac{2pt-qt-pq}{pqt}$ **26.** $\dfrac{az^2+2y^2}{xy^2z^2}$

B2.4, PAGES 167 – 168

2. $\dfrac{b}{2x}$ **4.** $\dfrac{10y}{z}$ **6.** $-\dfrac{3t}{2}$ **8.** $-\dfrac{49xu}{3}$ **10.** $\dfrac{q+2}{2}$ **12.** 1

14. $-a^2-2a+24$ **16.** $a+b$ **18.** $\dfrac{w}{4}$ **20.** $\dfrac{ay}{bx}$ **22.** $\dfrac{2}{a}$

24. $\dfrac{49b}{a}$ **26.** c^3 **28.** $\dfrac{1}{4aby}$ **30.** $\dfrac{10}{x^2}$ **32.** $\dfrac{25}{4}$

34. $\dfrac{2a+2b}{a-b}$ **36.** $x-2$

B2.5, PAGES 171 – 172

2. $\dfrac{2xy-3y}{x+3y}$ **4.** $-\dfrac{y}{x}$ **6.** $\dfrac{a+4b}{a-4b}$ **8.** $-\dfrac{x^2-3x}{x+3}$

10. $-\dfrac{y}{y^2+7}$ **12.** $\dfrac{a-b+16}{64}$ **14.** $\dfrac{a-5b}{2a+2b}$

16. ab **18.** $\dfrac{1-5a-5b}{a-b}$

B3.1, PAGES 185 – 186

2. 81   **4.** -125 **6.** 24   **8.** -64 **10.** 54   **12.** b^6 **14.** b^2c^5

16. $5u^2v^4$ **18.** $m^3 + 3n^3$ **20.** $-7a^3b - 5ab^2$ **22.** 5^{16} **24.** y^{11}

26. $-b^5$ **28.** $-n^8$ **30.** v^6 **32.** x^4y **34.** x^7y^3 **36.** x^{12}

38. $-x^{12}$

B3.2, PAGES 191 – 192

2. -100 **4.** -27 **6.** 5^{20} **8.** $8^4 9^3$ **10.** $-81y^4$ **12.** $81y^4$

14. $-72m^5$ **16.** $72t^9$ **18.** $-x^5$ **20.** b^8 **22.** xy^4z^9 **24.** $a^{12}b^8$

26. $-x^6$ **28.** m^9n^{18} **30.** x^6 **32.** $64a^{18}b^{12}c^6$ **34.** 25

36. u^4v^2 **38.** a^4 **40.** $-p^3q^3$

B3.3, PAGES 195 – 196

2. u^5v^5 **4.** x^2y^5 **6.** p^5q^2 **8.** a^4b^8 **10.** $72v^7w^8$ **12.** $c^{25}d^{21}$

14. $x^8y^7z^9$ **16.** $\dfrac{3r^3}{2}$ **18.** $2b^2$ **20.** $-u^2v^3$ **22.** p^3q^2 **24.** $\dfrac{1}{3x^4}$

26. $\dfrac{4}{w^5}$ **28.** $\dfrac{4s^4}{3r}$ **30.** $\dfrac{3a^3}{4b^5}$ **32.** $\dfrac{8x^9}{27y^6}$ **34.** $\dfrac{u^5}{v^7}$ **36.** $\dfrac{c^8}{d^2}$

38. $\dfrac{3wx^3}{8y^4}$

B3.4, PAGES 199 – 200

2. $\dfrac{1}{3}$ **4.** $\dfrac{1}{64}$ **6.** 1   **8.** $\dfrac{27}{64}$ **10.** N.D.   **12.** 1   **14.** N.D.

16. $\dfrac{1}{32}$ **18.** $\dfrac{m}{8}$ **20.** $\dfrac{b}{a^3}$ **22.** $\dfrac{1}{6}$ **24.** $-\dfrac{1}{3}$ **26.** $\dfrac{q^3}{2p^2}$

28. $\dfrac{1}{64c^2}$ **30.** d **32.** mn **34.** $\dfrac{b^6}{a^2}$ **36.** $\dfrac{9x^3}{y^4}$

38. $17a^2b^6$ **40.** $\dfrac{27s^6}{4r^8t^2}$

B3.5, PAGES 203 – 204

2. $\dfrac{2}{3}$ **4.** 9   **6.** $\dfrac{9}{16}$ **8.** $\dfrac{64}{81}$ **10.** $\dfrac{25}{81}$ **12.** 27   **14.** 1

16. m^2n^4 **18.** $\dfrac{b}{a}$ **20.** $\dfrac{p^6}{3^9q^6}$ **22.** $\dfrac{400}{a^6}$ **24.** $\dfrac{y^6}{81x^4}$ **26.** $\dfrac{s^{24}}{81}$

28. $\dfrac{1}{u^2v}$ **30.** $\dfrac{b^6c^3}{d^3}$ **32.** $\dfrac{1}{x^2}$ **34.** $\dfrac{1}{b^9c^{18}}$ **36.** $\dfrac{64a^6}{27}$ **38.** 1

40. $\dfrac{1}{xy^5}$

B3.6, PAGES 207 – 208

2. $-\dfrac{5}{3}$ **4.** 2   **6.** $\dfrac{a^2}{4b^3}$ **8.** $\dfrac{27}{2}$ **10.** $\dfrac{n^2}{8m}$ **12.** $\dfrac{y^4}{x^3}$ **14.** $\dfrac{s^3}{r}$

16. $\dfrac{a^2}{b}$ **18.** 25   **20.** $\dfrac{q^2}{p}$ **22.** $\dfrac{x^3z^2}{y^3}$ **24.** $\dfrac{1}{rs^3}$ **26.** $\dfrac{5}{6}$

28. $\dfrac{3a^4c^3}{b}$ **30.** $\dfrac{1}{(a+b)^2}$ **32.** $\dfrac{1}{a^2}+\dfrac{1}{b^2}$

B3.7, PAGES 211 – 212

2. 9.86×10^1 **4.** 1.76×10^3 **6.** 3.652422×10^2 **8.** 3.8×10^{-2}
10. 8.7×10^{-4} **12.** 4.14×10^{-6} **14.** 3.1416×10^6
16. 1.732×10^{-1} **18.** 2.5×10^{-1} **20.** 8.75×10^{-1}
22. 2.4×10^{-1} **24.** 8.75×10^1 **26.** 6.25×10^2 **28.** 7.29×10^2
30. 2.7182818   **32.** 1.414   **34.** 73,200,000   **36.** 3.36×10^2
38. 2.7×10^{-2} **40.** 2.25×10^4

B4.1, B4.2, PAGES 219 – 220

2. 9   **4.** a^4 **6.** $5x$ **8.** $12b^4$ **10.** $4x^3$ **12.** $5x^2\sqrt{2}$
14. $3x^3\sqrt{2}$ **16.** $4q^8\sqrt{2}$ **18.** $20w^5$ **20.** $5x^{25}$ **22.** $3a$
24. $5a^2$ **26.** $4x^3$ **28.** $2x$ **30.** $5z^3$ **32.** $\dfrac{5}{9}$ **34.** $3x^2$
36. $6x^6$ **38.** $3ab\sqrt{2}$ **40.** $x+y$ **42.** $4xy^2z^3$ **44.** $\dfrac{5y}{8}\sqrt{y}$

B4.3, B4.4, PAGES 223 – 224

2. 4   **4.** 6   **6.** $6a^2$ **8.** $6x^5$ **10.** $3x^4$ **12.** $5\sqrt{w}+w\sqrt{5}$
14. $\sqrt{ab}+b$ **16.** $\dfrac{2}{3}$ **18.** y **20.** 1   **22.** 3   **24.** 0
26. $2\sqrt{2}$ **28.** $3(\sqrt{2}-1)$ **30.** $x^2(1-\sqrt{x})$ **32.** $a(5-a)$
34. $\sqrt{2}(a+b)$ **36.** $2b(\sqrt{2}+\sqrt{7})$ **38.** $2(\sqrt{w}-\sqrt{z})$ **40.** a^2+b^2

B4.5, PAGES 227 – 228

2. $x=49$ **4.** $x=2$ **6.** $x=18$ **8.** $y=81$ **10.** $w=5$
12. $a=6$ **14.** $x=4$ **16.** $x=4$ **18.** $a=9$ **20.** $x=1$
22. $a=25$ **24.** $z=0$ **26.** $q=9$ **28.** $q=1{,}296$ **30.** $s=\dfrac{1}{2}$

B5.1, PAGES 231–232

2. $4t + 10 = 0$ **4.** $8y - 13 = 0$ **6.** $5m - 1 = 0$ **8.** $6s - 5 = 0$

10. $t = 2$ **12.** $r = -1$ **14.** $s = \dfrac{22}{5}$ **16.** $k = \dfrac{4}{9}$ **18.** $y = \dfrac{54}{13}$

20. $n = \dfrac{15}{7}$ **22.** $x = \dfrac{-10}{7}$ **24.** $s = \dfrac{19}{32}$ **26.** $y = 3$

28. 6 apples, 12 oranges, 12 pears   **30.** 16

B5.2, PAGES 235–236

2. $y = -\dfrac{1}{c}$ **4.** $s = \dfrac{2}{c}$ **6.** $x = -1$ **8.** $x = \dfrac{7c + 2}{4b}$

10. $F = 2C + 30$ **12.** $t = \dfrac{R}{Pr} - \dfrac{1}{r} = \dfrac{R - P}{Pr}$

14. $h = \dfrac{3S}{\pi r^2} - 3 = \dfrac{3S - 3\pi r^2}{\pi r^2}$ **16.** $b = \dfrac{3}{3a + 4c}$ **18.** $x = \dfrac{-5}{9r + 6}$

20. $a = \dfrac{2bc}{b + 3c}$ **22.** $s = \dfrac{12a}{1 - 2a}$

24. $y = \dfrac{a^2}{2a^2 - 2a - 2} = \dfrac{a^2}{2(a^2 - a - 1)}$ **26.** $n = \dfrac{4m}{5m - 11}$

B5.3, PAGES 239–240

2. $\dfrac{5}{4}$ **4.** $\dfrac{15}{16}$ **6.** $\dfrac{14}{30}$ **8.** yes, $\dfrac{111}{37} = \dfrac{3}{1}$ **10.** $x = 76$

12. 115.2 meters   **14.** 925 crew members   **16.** 116,640 times in a day
18. 695 deaths   **20.** \$1,130,038   **22.** $N = 3$ **24.** \$154   **26.** $x = 6$

B5.4, PAGES 245–246

2. $x = \dfrac{7}{3}$ **4.** $x = 9$ **6.** $t = -2$ **8.** $k = \dfrac{2}{3}$ **10.** $z = \dfrac{-3}{25}$

12. $s = \dfrac{25}{6}$ **14.** $x = \dfrac{60}{13}$ **16.** $t = 2$ **18.** $x = -6$ **20.** $b = 2$

22. $m = 4$ **24.** $r = -\dfrac{11}{14}$ **26.** 48 mph

B5.5, PAGES 249–250

2. $<$ **4.** $>$ **6.** $=$ **8.** $>$ **10.** $=$ **12.** $>$ **14.** $>$
16. $<$ **18.** $>$ **20.** $=$ **22.** $x < 2$ **24.** $x < 4$ **26.** $n > 2$

28. $t \leq 9$ **30.** $x \leq 3$ **32.** $T \leq -1$ **34.** $x \geq \dfrac{16}{7}$ **36.** $s > \dfrac{14}{13}$

38. $t > -\dfrac{1}{2}$ **40.** $y \leq -\dfrac{84}{5}$ **42.** $x > \dfrac{90}{41}$ **44.** $r < \dfrac{-125}{56}$

B5.6, PAGES 255–256

2. $a = 2$, $b = 1$ **4.** $x = 0$, $y = 2$ **6.** $u = -3$, $v = 4$

8. $p = 2$, $q = -5$ **10.** $s = \dfrac{3}{4}$, $t = 3$ **12.** $x = 2$, $y = \dfrac{1}{3}$

14. $u = \dfrac{16}{27}$, $v = \dfrac{-68}{27}$ **16.** $x = \dfrac{47}{22}$, $y = \dfrac{-1}{11}$

18. $r = \dfrac{-49}{23}$, $s = \dfrac{-30}{23}$ **20.** Mary Kay, 31; Jay, 26

22. 1,858 votes, 1,833 votes

B5.7, PAGES 259–260

2. $x = 3$, $y = 2$ **4.** $a = 4$, $b = -5$ **6.** $J = -3$, $K = 4$

8. $p = -3$, $q = -4$ **10.** $x = 12$, $y = -3$ **12.** $P = -2$, $Q = 5$

14. $a = \dfrac{4}{63}$, $b = \dfrac{-32}{21}$ **16.** $s = \dfrac{17}{71}$, $t = \dfrac{73}{71}$

18. $m = \dfrac{19}{11}$, $n = \dfrac{-2}{33}$ **20.** $x = \dfrac{11}{3}$, $y = \dfrac{-5}{6}$ **22.** $t = \dfrac{21}{38}$, $u = \dfrac{53}{38}$

24. $p = \dfrac{46}{11}$, $q = \dfrac{32}{11}$

B6.1, PAGES 269–270

2. $a = 2$, $a = -5$ **4.** $x = -\dfrac{3}{2}$, $x = \dfrac{5}{4}$ **6.** $q = -2$, $q = 2$

8. $p = -\dfrac{1}{3}$, $p = \dfrac{1}{3}$ **10.** $w = -\dfrac{1}{4}$, $w = 1$ **12.** $x = -2$

B6.2, PAGES 273–274

2. $x = 5, -5$ **4.** $b = 3, -2$ **6.** $x = \dfrac{1}{5}, \dfrac{-2}{5}$ **8.** $p = \dfrac{-2}{3}, \dfrac{-3}{2}$

10. $r = \dfrac{1}{3}, \dfrac{-2}{3}$ **12.** $a = -1$

B7.1, PAGES 277–278

2. G   **4.** H   **6.** J   **8.** J   **10.** D   **12.** F   **14.** H   **16.** I or J

18. P   **20.** F   **22.** C   **24.** S   **26.** I   **28.** I   **30.** Q   **32.** I

B7.2, PAGES 281 – 282

2. (number line, open circle at 5, shaded from 0 to 5, marks 0 1 2 3 4 5 6 7 8)

4. (number line, open circle at $\frac{3}{4}$, shaded from 0 to $\frac{3}{4}$, marks 0 $\frac{1}{8}$ $\frac{1}{4}$ $\frac{3}{8}$ $\frac{1}{2}$ $\frac{5}{8}$ $\frac{3}{4}$ $\frac{7}{8}$ 1)

6. (number line, open circles at 1 and 9, shaded between, marks 1 2 3 4 5 6 7 8 9 10 11)

8. (number line, open circles at −1 and 3, shaded between, marks −4 −3 −2 −1 0 1 2 3 4 5 6)

10. (number line, open circles at −0.1 and 0.6, shaded between, marks −0.2 −0.1 0 0.1 0.2 0.3 0.4 0.5 0.6 0.7)

12. $x > -3$ and $x < 1$ **14.** $x > -2$ and $x < 2$ **16.** $x > 2$ and $x < 4$

18. $x > -0.4$ and $x < -0.2$ **20.** $x > 1$ and $x < 3$

B7.3, PAGES 285 – 286

2. $\left(1\frac{1}{2}, 5\right)$ **4.** $(-4, 2)$ **6.** $(2, -6)$ **8.** $(-5, -5)$ **10.** $(4, -7)$

12. $\left(-1\frac{1}{2}, -4\right)$ **14.** $(-4, -7)$ **16.** $\left(1\frac{1}{2}, 1\right)$

18. -36.

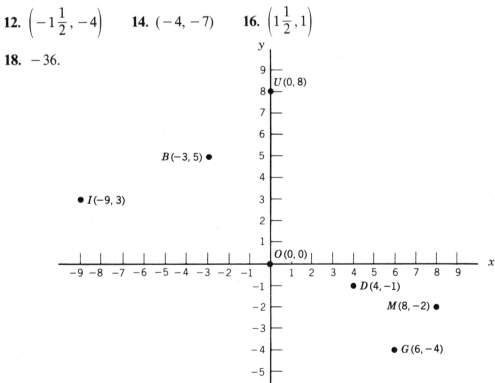

B7.4, PAGES 291–292

2.

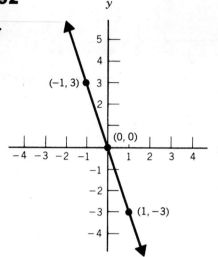

4.

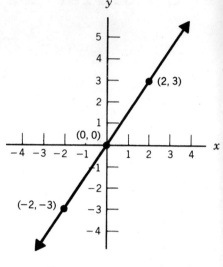

6. $y = -4x$

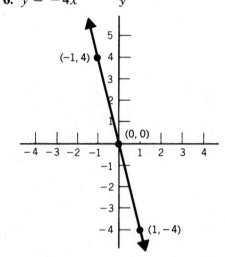

8. $y = \dfrac{-5}{2}x$

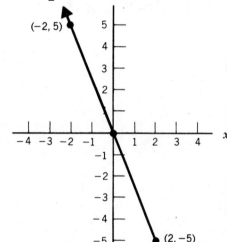

10. $y = -4$

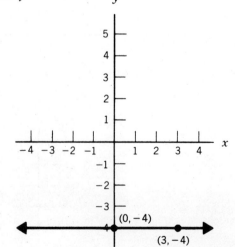

12. $x = -\dfrac{7}{3}$

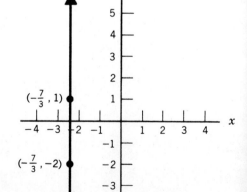

B7.5, PAGES 295 – 296

2. 2000   **4.** 51.9%   **6.** 59.3%   **8.** 65%   **10.** 14.4 hours
12. 34.5%   **14.** 27 men   **16.** 25 students   **18.** 27.8%
20. 43 men

C1.1, PAGES 309 – 310

Problems 1–5: Answers will vary
6. a. 9.40 cm   b. 46.81 cm   c. 120.65 cm   d. 39.69 cm
e. 248.92 cm   f. 19.53 cm
8. a. 43.25°   b. 187.05°   c. 176.7°   d. 217.22°

C1.2, PAGES 313 – 314

2. 30   **4.** 51   **6.** 10.8   **8.** 111.6   **10.** 18.6   **12.** $P = 4s$, 20
14. $P = a + b + c$, 13.8   **16.** $P = 6s$, 39.6   **18.** $P = 2l + 2w$, 51.6
20. $P = 2l + 2w$, 15.6

C1.3, PAGES 319 – 320

2. 4.5 square units   **4.** 12 square units   **6.** 442 square units
8. 31.71 square units   **10.** 300 square units   **12.** 24 square units
14. 6   **16.** width, 4; length 8

C2.1, PAGES 323 – 324

2. 32.656   **4.** 28.26   **6.** 22.6   **8.** 28   **10.** 12   **12.** 1 to 3, or $\frac{1}{3}$

14. 8.92

C2.2, PAGES 327 – 328

2. $R = 3$, $C = 18.84$, $A = 28.26$ **4.** $R = 6.69$, $D = 13.38$, $A = 140.53$
6. $D = 2$, $C = 6.28$, $A = 3.14$ **8.** $R = 95.54$, $D = 191.08$, $A = 28,661.58$
10. $R = 7$, $C = 43.96$, $A = 153.86$ **12.** 2   **14.** 25.42   **16.** 2

C3.1, PAGES 331 – 332

2. 0.875 cubic units   **4.** 0.000001 cubic units   **6.** 1 cubic unit
8. 216 cubic units   **10.** 12 cubic units   **12.** The volume will triple
14. 10 cubic units

C3.2, PAGES 335 – 336

2. 539.91π m^3 **4.** 121.5π cm^3 **6.** 36π in^3 **8.** 168.75π ft^3
10. eight times greater **12.** 8 inches

C4.1, PAGES 343 – 344

2. Answers may vary. **4.** $45°$ **6.** no
8. Hypotenuse is always twice as long as side b.

C4.2, PAGES 347 – 348

2. $115°$ **4.** $15°$ **6.** $\angle X = 52°$, $\angle Y = 32°$, $\angle Z = 96°$ **8.** $30°$
10. no **12.** $540°$

C5.1, PAGES 351 – 352

2. $\angle Y = 20°$, $\angle Z = 20°$ **4.** $\angle N = 55°$, $\angle O = 70°$
6. $\angle I = 50°$, $\angle J = 65°$ **8.** $\angle E = 22.5°$, $\angle F = 22.5°$
10. $\angle J = 15°$, $\angle K = 150°$

C5.2, PAGES 355 – 356

2. 20.4 **4.** 16 **6.** 5 or 3, 540 **8.** 2, 360

C6.1, PAGES 359 – 360

2. yes, ASA **4.** yes, ASA **6.** yes, SAS

C6.2, PAGES 363 – 364

2. XY; XZ **4.** $80°$ **6.** 1 **8.** 24 ft.

C7.1, PAGES 373 – 374

2. $39.0625, 36, 3.0625$ **4.** $9, 4, 5$ **6.** 6 **8.** $\sqrt{10}$ or 3.2
10. $5\sqrt{2}$ or 7.1 **12.** 2 **14.** $2\sqrt{313}$ or 35.4 **16.** $\sqrt{3.25}$ or 1.8
18. 18

C7.2, PAGES 379 – 380

2. 6.2   **4.** $4\sqrt{3}$ **6.** $24\sqrt{2} \approx 33.94$ **8.** 4 to $\sqrt{2}$, or $\dfrac{2\sqrt{2}}{1}$ **10.** $8\sqrt{2}$

12. 30°

C8.1, PAGES 383 – 384

2. Answers will vary somewhat   **4.** d and g; b and e **6.** 70°

8. 73°   **10.** $\angle y = \angle x$

C8.2, PAGES 387 – 388

2. a. $-\dfrac{5}{3}$ b. $\dfrac{3}{5}$ **4.** $l_1 \perp l_3$ **6.** yes   **8.** 135°

10. $\sqrt{74}$, $\sqrt{24} = 2\sqrt{6}$ **12.** $l_2 \| l_4$ **14.** 45°, 45°, 90°

16. $3^2 + 5^2 \neq 6^2$

ANSWERS TO ODD-NUMBERED PROBLEMS AND SELECTED SOLUTIONS TO ACHIEVEMENT TESTS

A1.1, PAGES 5 – 6

1. 4, 8, 2   **3.** 9, 4, 0   **5.** 3, 0, 0   **7.** 9, 7, 3, 2   **9.** 2, 0, 4, 0
11. 9, 9, 9, 9   **13.** 5, 0, 0, 0   **15.** 2, 8, 6, 3   **17.** 807   **19.** 6983
21. 8,462,900,473; eight billion four hundred sixty two million nine hundred thousand four hundred seventy three.
23. $(5 \times 10) + (8 \times 1)$
25. $(1 \times 1000) + (5 \times 100)$
27. $(3 \times 10,000) + (8 \times 100) + (9 \times 10)$
29. $(1 \times 100,000) + (5 \times 1000) + (9 \times 100)$
31. $(8 \times 1,000,000) + (9 \times 100,000) + (3 \times 1000)$
33. $(7 \times 1,000,000,000) + (9 \times 1,000,000) + (6 \times 100,000) + (3 \times 100)$

A1.2, PAGES 9 – 10

1. 552   **3.** 733   **5.** 1649   **7.** 443   **9.** 821   **11.** 842   **13.** 1011
15. 1051   **17.** 1516   **19.** 1552   **21.** 1276   **23.** 11,354
25. 15,555   **27.** 13,331   **29.** 90   **31.** 78   **33.** 259   **35.** 277
37. 1876   **39.** 755   **41.** 2889   **43.** 6174   **45.** 6174   **47.** 3676
49. 4443   **51.** 385   **53.** 939

A1.3, PAGES 13 – 14

1. 640   **3.** 684   **5.** 29,248   **7.** 38,640   **9.** 61,595   **11.** 38,724
13. 24,969   **15.** 25,170   **17.** 50,778   **19.** 74,175   **21.** 173,040
23. 412,888   **25.** 218,124   **27.** 85,865   **29.** 11   **31.** 39   **33.** 57
35. 521   **37.** 249r5   **39.** 87r3   **41.** 301r4   **43.** 113r95
45. 833r25   **47.** 65r9

A1.4, PAGES 17–18

1. +8 **3.** −1 **5.** −10 **7.** −26 **9.** −108 **11.** +50
13. −100 **15.** +192 **17.** −425 **19.** −311 **21.** −5
23. +8 **25.** +40 **27.** −66 **29.** +40 **31.** +106 **33.** −10
35. −200 **37.** −198 **39.** −175 **41.** +3 **43.** +42
45. −20 **47.** −14 **49.** +48 **51.** +90 **53.** −60
55. −5100 **57.** 0 **59.** +90 **61.** +3 **63.** +2 **65.** −2
67. −7 **69.** +4 **71.** +5 **73.** −5 **75.** −5 **77.** −3
79. +4

A1.5, PAGES 21–22

1. 71 **3.** 14 **5.** 106 **7.** 11 **9.** 49 **11.** 17 **13.** 17
15. 0 **17.** 34 **19.** 9 **21.** −54 **23.** 3 **25.** −13
27. −12 **29.** 7 **31.** 10 **33.** 33 **35.** 16 **37.** 40 **39.** 6
41. 1 **43.** 52 **45.** 495 **47.** 20 **49.** 324 **51.** 35 **53.** 8
55. 0 **57.** 98 **59.** 39

A1.6, PAGES 25–26

1. < **3.** < **5.** < **7.** < **9.** = **11.** is less than
13. is less than **15.** is greater than **17.** ..., −11, −10, −9
19. −5, −4, −3, −2, −1, 0, 1, ... **21.** ..., −9, −8, −7 **23.** 1, 2
25. 6 **27.** 336, 337, 338, 339, 340 **29.** −4, −2, 0, 3, 4
31. −20, −7, 9, 19, 23 **33.** 121, 101, 50, −16, −49
35. Answers will vary **37.** 2, 3, 4, 5 **39.** 0, 1, 2, 3, 4, 5, 6
41. There is none

ACHIEVEMENT TEST — A1, PAGES 27–30

1. B; 4 thousands, 7 hundreds, 3 tens, 0 ones
2. D; $(6 \times 100) + (0 \times 10) + (2 \times 1) = (6 \times 100) + (2 \times 1)$ **3.** B
4. E **5.** E **6.** C **7.** D **8.** E **9.** D; $8732 = (86 \times 101) + 46$
10. C; $2449 = (52 \times 47) + 5$ **11.** B
12. C; $(-16) - (-15) = (-16) + (+15) = -1$ **13.** D; same as 63×3
14. E; $(-3) \times (-9) = +27$
15. A; $42 - [16 + (9)3] = 42 - [16 + 27] = 42 - 43$
16. C; $4(6) - 5(4) = 24 - 20$ **17.** A; $6(12) + 8(7) = 72 + 56$
18. B; −452 is to the left of −375 **19.** A **20.** B

A2.1, PAGES 33–34

1. 4 **3.** 9 **5.** 14 **7.** 25 **9.** 39 **11.** 110 **13.** $\frac{11}{30}$

15. $2\frac{1}{20}$ **17.** $\frac{7}{24}$ **19.** $\frac{1}{3}$ **21.** $\frac{1}{3}$ **23.** $\frac{3}{2}$ **25.** $2\frac{1}{4}$

27. $2\frac{3}{12} = 2\frac{1}{4}$ **29.** $1\frac{12}{21} = 1\frac{4}{7}$ **31.** 3   **33.** 3   **35.** $1\frac{9}{45} = 1\frac{1}{5}$

37. $3\frac{14}{28} = 3\frac{1}{2}$ **39.** $4\frac{14}{21} = 4\frac{2}{3}$ **41.** $4\frac{35}{38}$ **43.** $\frac{11}{8}$ **45.** $\frac{79}{23}$

47. $\frac{89}{16}$ **49.** $\frac{37}{5}$ **51.** $\frac{28}{3}$ **53.** $\frac{104}{9}$ **55.** $\frac{121}{9}$ **57.** $\frac{159}{10}$

59. $\frac{249}{14}$ **61.** $\frac{218}{11}$ **63.** $\frac{328}{15}$ **65.** $\frac{356}{15}$ **67.** $\frac{440}{17}$ **69.** $\frac{557}{20}$

71. $\frac{746}{25}$

A2.2, PAGES 37 – 38

1. $\frac{9}{20}$ **3.** $\frac{5}{32}$ **5.** $\frac{20}{27}$ **7.** $\frac{6}{25}$ **9.** $\frac{1}{5}$ **11.** $\frac{3}{5}$ **13.** $\frac{1}{9}$

15. $\frac{2}{7}$ **17.** $\frac{5}{9}$ **19.** $-\frac{7}{12}$ **21.** $-\frac{5}{18}$ **23.** $\frac{1}{4}$ **25.** $3\frac{1}{8}$

27. 21   **29.** 5   **31.** 4   **33.** 14   **35.** 6   **37.** 10   **39.** 4

41. $6\frac{1}{4}$ **43.** $12\frac{3}{4}$ **45.** 15   **47.** $1\frac{1}{8}$ **49.** $\frac{10}{21}$ **51.** $\frac{3}{5}$

53. $-\frac{2}{3}$ **55.** $\frac{3}{11}$ **57.** $1\frac{2}{3}$ **59.** $8\frac{3}{4}$ **61.** $1\frac{1}{2}$ **63.** $2\frac{2}{5}$

65. $\frac{1}{2}$ **67.** $-5\frac{1}{3}$ **69.** $\frac{3}{10}$ **71.** $\frac{1}{14}$ **73.** $4\frac{1}{5}$ **75.** 6

77. $1\frac{37}{75}$ **79.** $\frac{2}{5}$ **81.** $\frac{5}{24}$ **83.** 4   **85.** $\frac{5}{21}$ **87.** 2

89. -8

A2.3, PAGES 41 – 42

1. $\frac{1}{2}$ **3.** $\frac{4}{5}$ **5.** $\frac{1}{4}$ **7.** $\frac{20}{21}$ **9.** $-\frac{4}{7}$ **11.** $-1\frac{1}{3}$

13. $-1\frac{11}{12}$ **15.** $\frac{4}{9}$ **17.** $3\frac{2}{5}$ **19.** $4\frac{4}{9}$ **21.** $\frac{5}{8}$

23. $3\frac{5}{8}$ **25.** $1\frac{1}{6}$ **27.** $-1\frac{1}{14}$ **29.** $\frac{19}{40}$ **31.** $\frac{7}{24}$ **33.** $\frac{5}{63}$

35. $1\frac{3}{10}$ **37.** $1\frac{1}{10}$ **39.** $\frac{17}{20}$ **41.** $2\frac{17}{24}$ **43.** $\frac{29}{56}$

45. $\frac{11}{72}$ **47.** $\frac{1}{42}$ **49.** $2\frac{1}{4}$ **51.** $-\frac{11}{18}$ **53.** $-\frac{7}{8}$ **55.** $2\frac{1}{24}$

57. $3\frac{13}{15}$ **59.** 7   **61.** $6\frac{43}{60}$ **63.** $-4\frac{8}{15}$ **65.** $1\frac{7}{132}$ **67.** $16\frac{7}{45}$

69. $12\frac{71}{180}$

A3.1, PAGES 45 – 46

1. 0.7   **3.** 7000   **5.** 8.4   **7.** 0.000302   **9.** 400,000.06
11. tenths   **13.** thousandths   **15.** hundred thousandths

17. thousands **19.** hundredths **21.** 0.002 **23.** 0.15 **25.** 0.8
27. 500.6 **29.** 16.18 **31.** 0.3 **33.** 0.8 **35.** 1.0 **37.** 21.0
39. 42.6 **41.** 0.17 **43.** 6.10 **45.** 13.00 **47.** 0.50 **49.** 0.637
51. 5.940 **53.** 11.000 **55.** 64.375 **57.** 0.091

A3.2, PAGES 49 – 50

1. 0.11 **3.** 0.99 **5.** 67.23 **7.** $10.56 **9.** 13.3 **11.** $0.44
13. 7.677 **15.** 46.401 **17.** 113.2105 **19.** 91.1140 **21.** $564.18
23. 15.0702 **25.** 0.22 **27.** 2.3 **29.** 50.52 **31.** $42.95
33. $3.39 **35.** $8.25 **37.** 43.81 **39.** 0.5386 **41.** 1.1773
43. 9.68 **45.** 8.984 **47.** 8.584

A3.3, PAGES 55 – 56

1. 1.05 **3.** 7.83 **5.** 0.00113 **7.** 0.08 **9.** 0.025 **11.** 0.96
13. 2.808 **15.** 0.52105 **17.** 0.001001 **19.** 0.6714 **21.** 0.415017
23. 0.27 **25.** $1.24 **27.** 6.3 **29.** 0.084 **31.** 3.096 **33.** 0.32
35. 0.35 **37.** 5.4 **39.** 125 **41.** 0.77 **43.** 500 **45.** 10.3
47. 7,125 **49.** 0.94 **51.** 12.54 **53.** 1.180 **55.** 1.4 **57.** 1.5

ACHIEVEMENT TEST — A2, A3, PAGES 57 – 60

1. C; an hour has 3 equal parts of 20 minutes each. Two of those equal parts is 40 minutes.

2. E; 16 days is 2 weeks and 2 days. 2 days is $\frac{2}{7}$ of a week. Hence, 16 days is $2\frac{2}{7}$ of a week.

3. B; $2\frac{45}{117} = 2\frac{45 \div 3}{117 \div 3}$ **4.** A; $7 \times 13 = 91$, $91 + 8 = 99$, $7\frac{8}{13} = \frac{99}{13}$

5. B; $\frac{5 + 11 - 7}{12} = \frac{9}{12} = \frac{9 \div 3}{12 \div 3}$

6. E; $\frac{31}{4} - \frac{21}{8} = \frac{62}{8} - \frac{21}{8} = \frac{41}{8} = 5\frac{1}{8}$

7. B; $\frac{23}{6} + \frac{7}{9} = \frac{69}{18} + \frac{14}{18} = \frac{83}{18} = 4\frac{11}{18}$

8. C; $\frac{200}{360} - \frac{168}{360} + \frac{165}{360} = \frac{197}{360}$

9. A; $\frac{14}{3} - \frac{7}{8} + \frac{1}{6} - \frac{13}{8} = \frac{112}{24} - \frac{21}{24} + \frac{4}{24} - \frac{39}{24}$

$= \frac{56}{24} = \frac{56 \div 8}{24 \div 8} = \frac{7}{3} = 2\frac{1}{3}$

10. E; $\frac{61}{9} + \frac{23}{5} + \frac{41}{15} - \frac{43}{10} = \frac{610}{90} + \frac{414}{90} + \frac{246}{90} - \frac{387}{90} = \frac{883}{90} = 9\frac{73}{90}$

11. A; $\dfrac{7}{9 \times 5} = \dfrac{7}{45}$ **12.** C; $\dfrac{\overset{3}{\cancel{33}}}{\cancel{7}_{1}} \times \dfrac{\overset{4}{\cancel{28}}}{\cancel{11}_{1}} = 3 \times 4 = 12$

13. A; $\dfrac{2}{3} \times \dfrac{\overset{2}{\cancel{8}}}{1} \times \dfrac{5}{\cancel{4}_{1}} = \dfrac{2 \times 2 \times 5}{3 \times 1} = \dfrac{20}{3} = 6\dfrac{2}{3}$

14. E; $\dfrac{39}{8} \div \dfrac{7}{4} = \dfrac{39}{\cancel{8}_{2}} \times \dfrac{\overset{1}{\cancel{4}}}{7} = \dfrac{39}{14} = 2\dfrac{11}{14}$

15. D; $\dfrac{\overset{2}{\cancel{12}}}{35} \times \dfrac{1}{\cancel{30}_{5}} = \dfrac{2}{35 \times 5} = \dfrac{2}{175}$

16. A; $36 \div \dfrac{47}{8} = \dfrac{36}{1} \times \dfrac{8}{47} = \dfrac{288}{47} = 6\dfrac{6}{47}$

17. C; the left-most 9 is to be rounded. The digit to its right is less than 5. Hence, the answer is 0.9.

18. C; the 7 is to be rounded. The digit to its right is less than 5. Hence, the answer is 0.07.

19. B; 5.55⑤5, the circled 5 is to be rounded. The digit to its right is 5 or greater. Hence, the answer is 5.556.

20. B   **21.** E   **22.** A

A4.1, PAGES 63 – 64

1. 27   **3.** 25   **5.** 64   **7.** 256   **9.** 100   **11.** 100,000,000

13. 216   **15.** 16,807   **17.** $\dfrac{8}{27}$ **19.** $\dfrac{25}{64}$ **21.** 0.36   **23.** 1.96

25. $\dfrac{5}{32}$ **27.** $\dfrac{12}{49}$ **29.** 2^{15} or 32,768   **31.** 3^9 or 19,683

33. 10^8 or 100,000,000   **35.** 12^4 or 20,736   **37.** 6^5 or 7,776

39. 27   **41.** 1   **43.** 225   **45.** 113   **47.** 793   **49.** 52   **51.** $\dfrac{40}{81}$

A4.2, PAGES 67 – 68

1. 9   **3.** 6   **5.** 8   **7.** 13   **9.** 18   **11.** 12   **13.** 27   **15.** 100

17. 0.3   **19.** 0.02   **21.** $\dfrac{5}{6}$ **23.** $\dfrac{2}{9}$ **25.** $\dfrac{4}{3}$ **27.** 2, 3

29. 4, 5   **31.** 8, 9   **33.** 11, 12   **35.** 24, 25   **37.** 22, 23

39. 2, 3   **41.** 28, 29   **43.** 2, 3   **45.** 22, 23   **47.** 3, 4

49. 316, 317

A5.1, PAGES 73 – 74

1. 3.5   **3.** 3.72   **5.** 1.34   **7.** 1.2   **9.** 3.25   **11.** 0.55   **13.** 4.5
15. 1.36   **17.** 2.25   **19.** 0.375   **21.** 0.012   **23.** 0.175
25. 3.625   **27.** 1.75   **29.** 11.3125   **31.** 9.35   **33.** 8.55   **35.** 0.75
37. 0.625   **39.** 0.028   **41.** 0.32   **43.** 0.1875   **45.** 0.00096
47. 0.15   **49.** 0.125   **51.** $0.49\overline{3}$ **53.** 0.17   **55.** 1.18   **57.** 1.36
59. 14.12   **61.** 1.37   **63.** 0.69   **65.** $7.80\overline{285714}$ **67.** 0.75
69. 1.5

A5.2, PAGES 77 – 78

1. $\dfrac{1}{5}$ **3.** $\dfrac{3}{20}$ **5.** $\dfrac{32}{25}$ **7.** $\dfrac{201}{200}$ **9.** $\dfrac{64}{25}$ **11.** $\dfrac{5}{16}$ **13.** $\dfrac{3}{8}$

15. $\dfrac{7}{8}$ **17.** $\dfrac{28}{25}$ **19.** $\dfrac{737}{25,000}$ **21.** $\dfrac{3003}{1250}$ **23.** $\dfrac{170,717}{2000}$

25. $\dfrac{9}{2500}$ **27.** $\dfrac{7}{64}$ **29.** $\dfrac{23}{160}$ **31.** $\dfrac{977}{80}$ **33.** $\dfrac{15,432}{125}$

35. $\dfrac{876,543}{1000}$ **37.** $\dfrac{112}{15,625}$ **39.** $\dfrac{5}{128}$ **41.** $\dfrac{1}{31,250}$ **43.** $\dfrac{2,000,001}{78,125}$

45. $1\dfrac{1}{4}$ **47.** $1\dfrac{1}{125}$ **49.** $6\dfrac{1}{4}$ **51.** $1\dfrac{7}{25}$ **53.** $65\dfrac{21}{200}$

55. $122\dfrac{481}{2500}$ **57.** $15\dfrac{903}{2000}$ **59.** $48\dfrac{421}{2500}$ **61.** $\dfrac{1}{4}$, 0.25   **63.** $\dfrac{5}{4}$, 1.25

A6.1, PAGES 81 – 82

1. 31°   **3.** 244 dictionaries   **5.** 78   **7.** 60.5 km/hr   **9.** 78

A6.2, PAGES 85 – 88

1. 345%   **3.** $\dfrac{69}{20,000}$ **5.** 0.0345   **7.** 25%   **9.** $\dfrac{1}{4}$ **11.** 0.0025

13. 86%   **15.** 50%   **17.** 75%   **19.** 2%   **21.** 412 students
23. 20%   **25.** $10.50   **27.** 42 inches   **29.** $83\dfrac{1}{3}$%   **31.** 60%
33. 40

A6.3, PAGES 91 – 94

1. 58.2 gallons
3. 269 hours   **5.** 1.8 yards   **7.** 20,894 gallons   **9.** 4:06 p.m.
11. 12:45 p.m.   **13.** $154.50

ACHIEVEMENT TEST — A4, A5, A6, PAGES 95 – 98

1. D; $7 \times 7 = 49$ **2.** A; $8 \times 8 = 64$ **3.** C; $7^2 = 49$

4. D; $\frac{4}{25} + \frac{49}{25} = \frac{53}{25}$ **5.** B; $7^2 = 49$ and $8^2 = 64$

6. C; $0.4 \times 0.4 = 0.16$ **7.** C; $7 \div 8 = 0.875$ **8.** A; $1.2 + 2.66 = 3.86$

9. E; $5 + 0.7 + 0.003 = 5.703$ **10.** C; $\frac{455 \div 5}{10,000 \div 5} = \frac{91}{2000}$

11. A; $\frac{9}{40} \div \frac{0.05}{1} = \frac{9}{40} \times \frac{1}{0.05} = \frac{9}{2} = 4.5$ **12.** B; $246 \div 4 = 61\frac{1}{2}$

13. B; $(83 + 92 + 66 + 54 + 85 + 73) \div 6 = 453 \div 6 = 75.5$

14. D; cost of one pound $= \frac{\$3.29}{7} = \$0.47$, cost of 12 pounds $= \$0.47 \times 12$
$= \$5.64$.

15. A; cost of 1 kg $= \frac{\$4.24}{8} = \$0.53$, cost of 3 kg $= \$0.53 \times 3 = \$1.59$.

16. C; 15 min. per 1b. $\times$ 8 lbs. $= 120$ min $= 2$ hrs. 2 hours before 8:00 p.m. is
6:00 p.m.

17. D; $\frac{63}{90} = 0.7 = 70\%$ **18.** C; $\frac{35}{70} = 0.5 = 50\%$

19. D; $\frac{15}{60} = 0.25 = 25\%$ **20.** A; $\frac{90}{60} = 1.5 = 150\%$

B1.1, PAGES 103 – 104

1. $5 + t$ or $t + 5$ **3.** $-4x$ **4.** $9k$ **5.** $-2rs$ **6.** $5pq$
7. $(-7)(-x)y$ or $7xy$ **9.** $4u + (-5v)$ **11.** no, yes
13. yes, yes   **15.** no, yes   **17.** yes, yes   **19.** $14x$ **21.** $-2y$
23. $4m$ **25.** $-5s$ **27.** $4x - 2y$ **29.** $5 - 3m$ or $-3m + 5$
31. $4k - 4a$ or $-4a + 4k$ **33.** $9s - t$ **35.** $-3a - 2b$ **37.** $22rt + s$
39. $8a + 3b$ **41.** $-15cd + 6c$ or $6c - 15cd$ **43.** $-3b$
45. $-\frac{1}{3}s + t$ or $t - \frac{1}{3}s$ **47.** $\frac{13}{12}a + \frac{11}{6}b$ **49.** $6abc - 4ab + a$

B1.2, PAGES 107 – 108

1. -5 **3.** 4   **5.** 5   **7.** 8   **9.** 27   **11.** 5   **13.** -16 **15.** 0
17. 7   **19.** 60   **21.** -20 **23.** 15   **25.** -9 **27.** -10
29. -16 **31.** -19 **33.** -206 **35.** 3   **37.** 5   **39.** 38
41. -79 **43.** 9   **45.** 21   **47.** 21

B1.3, PAGES 111 – 112

1. $8x^2 - x + 9$ **3.** $9x^2 + 3x + 2$ **5.** $6y^2 + 3y + 2$
7. $3u^2 - 3u - 4$ **9.** $3v^2 - 1$ **11.** $7w^2 - w$ **13.** 2   **15.** 0
17. $3x^2 - 7x + 5$ **19.** $3x^2 - 6x - 9$ **21.** $-2t^2 - 9t + 11$

23. $2r^2 + 2r$ **25.** $3a^2 + a + 2$

27. $-5m^2 - 2m + 2$ **29.** $7x^2 - 7x + 19$ **31.** $-9t^2 + 6t - 9$

33. $2k^3 - 6k^2 + 5k + 12$

35. $-2u^3 + 4u^2 + 15u - 4$ **37.** $-6v^3 - 9v^2 - 11v + 20$

39. $-7x - 12$

B1.4, PAGES 115 – 116

1. x^4 **3.** b^4 **5.** $12u^3v$ **7.** $-12a^3b$ **9.** $28c^5d^3$ **11.** 288

13. -648 **15.** 81 **17.** $3s^4 + 2s^3$ **19.** $3n^3 - 6n$ **21.** $10s^3 - 15s^2$

23. $18m^4 - 12m^3$ **25.** $2c^3xy + 8c^3y$ **27.** $27a^6 - 18a^5 + 36a^4$

29. $-14p^3q^3 + 28p^2q^2 + 14pq^2$ **31.** $-20g^5h^3 - 10g^4h^2 + 5g^3h^3$

33. $48a^4b^3c^3 + 40a^3b^5c + 32a^3b^2c^3$ **35.** $-30u^7v^4 + 42u^5v^5 - 54u^4v^3w^2$

37. $64x^9y^9z^5 + 72x^5y^5z^2 - 56x^4y^7z^2 + 24x^9y^5z^2$

39. $36u^{12}v^{11}w^2 - 45u^{14}v^8w^3 - 54u^7v^{16} + 51u^7v^8w^{10}$

B1.5, PAGES 121 – 122

1. $x^2 + 7x + 12$ **3.** $2y^2 + 19y + 42$ **5.** $2a^2 - 5ab - 3b^2$

7. $6p^2 - 13pq + 6q^2$ **9.** $16b^2 - c^2$ **11.** $15x^2 + 37xy + 20y^2$

13. $-4s^2 + 12st - 9t^2$ **15.** $6a^2b^2 - abc - 2c^2$ **17.** $20 + 7n - 6n^2$

19. $4v^2 - w^2$ **21.** $49x^2 - 4y^2$ **23.** $25m^2 - 36n^2$ **25.** $25ab$

27. $a^4 - b^2$ **29.** $6x^2 + 3xy + 2xz + yz$ **31.** $2r^3 + 2r^2s - 3rs^2 - 3s^3$

33. $7u^2v^2 + 20uvw - 3w^2$ **35.** $9r^4s^2 - 4t^4$

37. $24w^4x^2 - 16wx^5 + 27w^5x - 18w^2x^4$

39. $30a^5bc^3 - 25a^4b^3c^2 + 18a^2b^3c^2 - 15ab^5c$

B1.6, PAGES 125 – 126

1. $r^2 + 4r + 4$ **3.** $t^2 - 8t + 16$ **5.** $9 - 6s + s^2$ **7.** $x^2 + 2xy + y^2$

9. $4a^2 - 20ab + 25b^2$ **11.** $4r^2s^2 - 20rs + 25$

13. $64a^4 + 16a^2b + b^2$ **15.** $4s^2 + 12st + 9t^2$

17. $16p^2q^2 + 24pqr + 9r^2$ **19.** $\dfrac{1}{9} + 2x + 9x^2$ **21.** $\dfrac{1}{4} - 2r + 4r^2$

23. $a^2 + 2 + \dfrac{1}{a^2}$ **25.** $16m^4n^2 - 24m^3n + 9m^2$

27. $36m^6 - 12m^5n + m^4n^2$ **29.** $4u^4v^6 + 20u^2v^5w + 25v^4w^2$

31. $49a^2b^4c^6 - 112a^4b^4c^4 + 64a^6b^4c^2$ **33.** $25j^2 + \dfrac{15jkv}{2} + \dfrac{9k^2v^2}{16}$

35. $12\dfrac{1}{4}$ **37.** $90\dfrac{1}{4}$ **39.** $2x^2 + 2y^2$

B1.7, PAGES 129 – 130

1. $4x + 3$ **3.** $5t - 4$ **5.** $13m^2 - 3m$ **7.** $-42p^4 + 6p^3 - 2$

9. $28s^2 - 26s$ **11.** $3z^4 + 5z^2$ **13.** $-12u^2 - 11u$

15. $-12q^3 + 15q$ **17.** $4y^2 - 3y + 2$ **19.** $-4n^4 - n^2 - 3$
21. $-17v^2 + v - 16$ **23.** $39m^3 - 24m^2 - 2$
25. $-6t^3 - \dfrac{1}{2}t^2 + 2t$ **27.** $6x^3 - \dfrac{1}{2}x^2 + \dfrac{2}{3}x + 1$

B1.8, PAGES 133 – 134

1. $x + 1$ **3.** $3x - 1$ **5.** $x + 1$ **7.** $x - 3$ **9.** $3x^2 + 6x + 2$
11. $x^3 - 2$ **13.** $27x^3 + 9x^2 + 3x + 1$ **15.** $x^3 - 3x^2 + 3x - 1$
17. $x^2 + 2x + 1$ **19.** $x^4 + x^3 + x^2 + x + 1$ **21.** $x^3 + x^2 + x + 1$

B1.9, PAGES 137 – 138

1. $5(x - 2)$ **3.** $6(1 + p^2)$ **5.** $2(r - st)$ **7.** $3a(3a + 16b)$
9. $19d(3c - 2d)$ **11.** $m(3m^2 + m + 3)$ **13.** $6pq(2p^2 - 3p + 4q^2)$
15. $2c^2(c^3 + 2c^2 + 3c + 4)$ **17.** $2r(2s + 3r - 4s^2)$
19. $7a^3b^3c^2(14a^2c^2 - 6b^3 - 9abc)$ **21.** $(x + y)(u - v)$
23. $(a + 2)(2 + r)$ **25.** $(c + 4)(2d - 3)$ **27.** $(a + b)(x - y)$
29. $(b + 1)(ab + 1)$ **31.** $(m + 6)(m - 1)$ **33.** $(4c + 1)(4c + d)$
35. $(s^2 - 3)(t^2 + 1)$ **37.** $(2ac - 3b)(4ab - 5)$

B1.10, PAGES 141 – 142

1. $(x + 2)(x + 1)$ **3.** $(x + 1)^2$ **5.** $(3x + 1)^2$ **7.** N.F.
9. $(x - 3)(x - 2)$ **11.** N.F.   **13.** $(2x - 3)(x + 5)$ **15.** $(x - 6)^2$
17. $(x - 14)(x + 3)$ **19.** $(4x - 7)^2$ **21.** $(3x + 1)(x - 2)$
23. $5(x + 1)^2$ **25.** $(3x - 2)(2x - 1)$ **27.** N.F.
29. $(7x - 2)(3x + 1)$ **31.** N.F.   **33.** N.F.   **35.** $(4x - 3)(3x + 5)$
37. $4(2x - 3)(2x + 1)$ **39.** $(3x + 7)(3x - 2)$

B1.11, PAGES 145 – 146

1. b, b **3.** $2, 2, 3$ **5.** $9, 6, 9, 6$ **7.** $y, x, 3, 3$ **9.** a^3, b^4, a^3, b^4
11. $3, 3, 1, 3, 1$ **13.** $3, 1, 2, 1, 2$ **15.** $16(x - 3y)(x + 3y)$
17. N.F.   **19.** $12y(y - 4)(y + 4)$ **21.** N.F.   **23.** $2ab(2a - 3b)(2a + 3b)$
25. $(ab - 3)(ab + 3)$ **27.** $(4 - 3)(4 + 3)(16 + 9)$ **29.** N.F.
31. N.F.   **33.** $(8y - 7x)(8y + 7x)$ **35.** $2z(z^2 - 3)(z^2 + 3)$
37. $pq(x - 2qy)(x + 2qy)$ **39.** 1599 **41.** 9999 **43.** 3584
45. 119 **47.** 725

ACHIEVEMENT TEST — B1, PAGES 147 – 150

1. B; $(12 - 3 + 1)x = 10x$ **2.** D; $-4a + a + 13b - 6b = -3a + 7b$
3. D; $3^2 - 2(-6)(3) - 3 = 9 - (-36) - 3 = 9 + 36 - 3$

4. C; $7x^2 - 8x - 5 - 10x^2 - 4x + 7 = \boxed{7x^2 - 10x^2} - \underline{8x - 4x} - 5 + 7$

5. E; $4a^2b^2(3ab) - 4a^2b^2(5ab^3) - 4a^2b^2(8a^3b)$

6. A; $(5p - 3q)(3p) - (5p - 3q)(4q) = (5p)(3p) - (3q)(3p) - (5p)(4q) - (-3q)(4q) = 15p^2 - 9pq - 20pq + 12q^2$

7. D; $7(2) + 7(4y) - 3x(2) - 3x(4y) = 14 + 28y - 6x - 12xy$

8. B; $6(5r) + 6(6) - 5r(5r) - 5r(6) = 30r + 36 - 25r^2 - 30r$

9. C; $3(3) - 3\left(\dfrac{2t}{3}\right) - \dfrac{2t}{3}(3) - \left(-\dfrac{2t}{3}\right)\left(\dfrac{2t}{3}\right) = 9 - 2t - 2t + \dfrac{4t^2}{9}$

10. E; $(2m)^2 - 2mn^2 - 2mn^2 + (n^2)^2$ **11.** A   **12.** C   **13.** A

14. B   **15.** A   **16.** D   **17.** B   **18.** E   **19.** C

20. D; $\left(9 + \dfrac{1}{2}\right)^2 = 81 + 9 + \dfrac{1}{4}$

B2.1, PAGES 155–156

1. $\dfrac{1}{2}$ **3.** $-\dfrac{1}{6}$ **5.** $\dfrac{3b}{4}$ **7.** $\dfrac{2t}{3}$ **9.** 1   **11.** $-\dfrac{2a}{3}$

13. $\dfrac{-4b}{5}$ **15.** $\dfrac{7x^2}{8}$ **17.** $-3q^2$ **19.** $\dfrac{4}{5y}$ **21.** $\dfrac{1}{a+b}$

23. $\dfrac{3}{4}$ **25.** $p - 4$ **27.** -1 **29.** $\dfrac{x+y}{x-y}$

B2.2, PAGES 159–160

1. 45   **3.** 108   **5.** 9   **7.** 10   **9.** 11   **11.** -20 **13.** $1\dfrac{1}{3}$

15. $-\dfrac{1}{2}$ **17.** 8   **19.** $\dfrac{1}{8}$

B2.3, PAGES 163–164

1. $\dfrac{5}{b}$ **3.** $-\dfrac{3x}{y}$ **5.** $\dfrac{x}{4y}$ **7.** 4   **9.** $\dfrac{64 - 2a^2b^2}{8ab} = \dfrac{32 - a^2b^2}{4ab}$

11. $\dfrac{3y - 5yb}{ab}$ **13.** $\dfrac{a^2 - b^2}{ab}$ **15.** $\dfrac{x^2 + 3x + 5y}{xy}$ **17.** $\dfrac{-8x}{x^2 - 4}$

19. $\dfrac{7x + 1}{x^2 + x - 6}$ **21.** $-a - 4$ **23.** $\dfrac{16az + 8bx}{xyz}$ **25.** $\dfrac{5}{x^2 - x - 6}$

B2.4, PAGES 167–168

1. $\dfrac{1}{4}$ **3.** 1   **5.** $\dfrac{ax^2}{6}$ **7.** $\dfrac{a^3b}{cd}$ **9.** $\dfrac{x+3}{6}$ **11.** $\dfrac{2-y}{4}$

13. $\dfrac{6a + 9}{a + 3}$ **15.** $\dfrac{4y}{a + b}$ **17.** $\dfrac{a^2 - b^2}{2}$ **19.** 6   **21.** $\dfrac{3}{2}$

23. -4 **25.** $\dfrac{25pq^2}{2}$ **27.** $a - b$ **29.** $50a + 100b$ **31.** $\dfrac{9xy}{4x + 4y}$

33. $\dfrac{a(b + c)^2}{a + b}$ or $\dfrac{ab^2 + 2abc + ac^2}{a + b}$ **35.** $4a^2 - 4b^2$

B2.5, PAGES 171 – 172

1. $\dfrac{a}{a + 1}$ **3.** $\dfrac{5b - c}{5a + b}$ **5.** $\dfrac{4b}{4b - 3}$ **7.** $\dfrac{2 - x + y}{y^3}$ **9.** $\dfrac{a^2 - b^2}{a - bx}$

11. $\dfrac{2a + 5}{a^2 + 5}$ **13.** $\dfrac{x^2 - y^2 - 6x - 6y}{x - y}$ **15.** $\dfrac{y^2 - x^2}{y^2 + x^2}$

17. $\dfrac{9y^2 - 4x^2}{x}$

ACHIEVEMENT TEST — B2, PAGES 173 – 180

1. D; $\dfrac{6n \div 6n}{12n \div 6n} = \dfrac{1}{2}$ **2.** D; $\dfrac{8x \div 4x}{12xy \div 4x} = \dfrac{2}{3y}$

3. E; $\dfrac{45abc \div 15ac}{60ac \div 15ac} = \dfrac{3b}{4}$ **4.** B; $\dfrac{-4pq^2 \div 2pq}{2p^2q \div 2pq} = \dfrac{-2q}{p}$

5. D; $\dfrac{4p^2 \div 4p}{4p(2 + q) \div 4p} = \dfrac{p}{2 + q}$

6. B; $\dfrac{2(a + b)(a - b) \div (a - b)}{-1(a - b) \div (a - b)} = \dfrac{2(a + b)}{-1}$

7. E; $\dfrac{(x + y)(x - y)}{(x + y)(x - y)} = 1$

8. A; $\dfrac{a}{b} = \dfrac{3}{-2}$ **9.** D; $2p^2 = 2\left(-\dfrac{1}{3}\right)^2 = 2\left(\dfrac{1}{9}\right)$

10. A; $\dfrac{xy}{x(1 - y)} = \dfrac{y}{1 - y} = \dfrac{-6}{1 - (-6)} = \dfrac{-6}{7}$

11. E; $\dfrac{2^2 + (-3)^2}{(2 + (-3))^2} = \dfrac{4 + 9}{(-1)^2} = \dfrac{13}{1} = 13$

12. D; $\dfrac{-2(-3)^2}{((-2) - (-3))^2} = \dfrac{-2(9)}{(-2 + 3)^2} = \dfrac{-18}{1^2} = -18$

13. B; $\dfrac{(-4)^2 + (-4)^2}{-4 + -4} = \dfrac{16 + 16}{-18} = \dfrac{32}{-8} = -4$

14. B; $\dfrac{x^2}{x} - \dfrac{1}{x} = \dfrac{x^2 - 1}{x}$ **15.** B; $\dfrac{1}{y} - \dfrac{xy}{y} = \dfrac{1 - xy}{y}$

16. D; $\dfrac{2}{x + 1} \cdot \dfrac{(x - 1)}{(x - 1)} + \dfrac{1}{x - 1} \cdot \dfrac{(x + 1)}{(x + 1)} = \dfrac{2(x - 1) + (x + 1)}{(x + 1)(x - 1)}$

$= \dfrac{2x - 2 + x + 1}{x^2 - 1}$

17. A; $\dfrac{a-b}{xy} \cdot \dfrac{z}{z} + \dfrac{c-b}{xz} \cdot \dfrac{y}{y}$

18. C; $\dfrac{a^2}{a-3} + \dfrac{9}{-1(a-3)} = \dfrac{a^2}{a-3} + \dfrac{-9}{a-3} = \dfrac{a^2-9}{a-3} = \dfrac{(a-3)(a+3)}{a-3}$

19. B; $\dfrac{x^2}{x-y} + \dfrac{y^2}{-1(x-y)} = \dfrac{x^2}{x-y} - \dfrac{y^2}{x-y} = \dfrac{x^2-y^2}{x-y}$

$\qquad = \dfrac{(x-y)(x+y)}{x-y}$

20. B; $\dfrac{(a-b)c-(b-c)a}{abc} = \dfrac{ac-bc-ab+ac}{abc}$

21. B; $\dfrac{(x+y)(x+y)(x-y)(x-y)}{(x+y)(x-y)} = (x+y)(x-y)$

22. C; $\dfrac{(a+3)(a-3)(a+3)}{(a-3)(a+3)}$ **23.** C; $\dfrac{(x+1)(x+1)(x-1)}{(x+1)(x-1)(x+1)}$

24. D; $\dfrac{x}{x+1} \cdot \dfrac{y}{x}$ **25.** D; $\dfrac{a+b}{a-b} \cdot \dfrac{(a-b)(a+b)}{(a+b)(a+b)}$

26. A; $\dfrac{a^2 y}{b} \div \dfrac{ay}{b^2} = \dfrac{a^2 y}{b} \cdot \dfrac{b^2}{ay} = \dfrac{ayb \cdot ab}{ayb}$

27. B; $\dfrac{a+b}{xy} \div \dfrac{xy}{a-b} = \dfrac{a+b}{xy} \cdot \dfrac{a-b}{xy}$

28. A; $\dfrac{a}{\dfrac{5a}{5} - \dfrac{1}{5}} = \dfrac{a}{\dfrac{5a-1}{5}} = \dfrac{a}{1} \cdot \dfrac{5}{5a-1}$

29. D; $\dfrac{\dfrac{2a}{3}}{\dfrac{3a}{3} - \dfrac{1}{3}} = \dfrac{\dfrac{2a}{3}}{\dfrac{3a-1}{3}} = \dfrac{2a}{3} \div \dfrac{3a-1}{3} = \dfrac{2a}{3} \cdot \dfrac{3}{3a-1}$

30. C; $\dfrac{\dfrac{n \cdot n}{n} + \dfrac{3}{n}}{\dfrac{1 \cdot n}{n} + \dfrac{3}{n}} = \dfrac{\dfrac{n^2+3}{n}}{\dfrac{n+3}{n}} = \dfrac{n^2+3}{n} \div \dfrac{n+3}{n} = \dfrac{n^2+3}{n} \cdot \dfrac{n}{n+3}$

31. A; $\dfrac{\dfrac{x \cdot x}{x} + \dfrac{1}{x}}{\dfrac{x \cdot x}{x} - \dfrac{1}{x}} = \dfrac{\dfrac{x^2+1}{x}}{\dfrac{x^2-1}{x}} = \dfrac{x^2+1}{x} \div \dfrac{x^2-1}{x} \div \dfrac{x^2+1}{x} \cdot \dfrac{x}{x^2-1}$

32. C; $\dfrac{\dfrac{t^2-4}{t}}{\dfrac{4-t^2}{t^2}} = \dfrac{t^2-4}{t} \div \dfrac{4-t^2}{t^2} = \dfrac{t^2-4}{t} \cdot \dfrac{t^2}{4-t^2} = \dfrac{t^2(t^2-4)}{(-1)t(t^2-4)}$

33. D; $\dfrac{\dfrac{4x^2-9}{6x}}{\dfrac{3+2x}{3}} = \dfrac{4x^2-9}{6} \div \dfrac{3+2x}{3} = \dfrac{4x^2-9}{6} \cdot \dfrac{3}{3+2x}$

$\qquad = \dfrac{(2x-3)(2x+3)}{6} \cdot \dfrac{3}{2x+3} = \dfrac{2x-3}{2} = \dfrac{2x}{2} - \dfrac{3}{2} = x - \dfrac{3}{2}$

34. A; $\dfrac{\dfrac{5(5-r)-(r-1)}{(r-1)(5-r)}}{\dfrac{2}{r-5}} = \dfrac{26-6r}{(r-1)(5-r)} \div \dfrac{2}{r-5}$

$= \dfrac{2(13-3r)}{(r-1)(-1)(r-5)} \cdot \dfrac{r-5}{2} = -\dfrac{13-3r}{r-1} = \dfrac{3r-13}{r-1}$

35. A; $\dfrac{3a}{5} \div \dfrac{a^2-3a}{45-5a^2} = \dfrac{3a}{5} \cdot \dfrac{45-5a^2}{a^2-3a} = \dfrac{3a}{5} \cdot \dfrac{5(3-a)(3+a)}{a(a-3)}$

$= \dfrac{3a}{5} \cdot \dfrac{5(-1)(a-3)(a+3)}{a(a-3)} = (-1)(3)(a+3) = -3a-9$

B3.1, PAGES 185 – 186

1. 32   **3.** -64 **5.** 36   **7.** -125 **9.** 24   **11.** a^5 **13.** a^3b^4
15. $3x^3y^2$ **17.** $7p^4-q^3$ **19.** $-3x^2y+13xy^2$ **21.** 7^{13} **23.** x^{10}
25. $-a^6$ **27.** m^6 **29.** u^9 **31.** $-a^4b^3$ **33.** a^6b^3 **35.** s^6
37. $-s^6$

B3.2, PAGES 191 – 192

1. 36   **3.** -16 **5.** $6^5 \cdot 7^5$ **7.** 13^{12} **9.** $4x^2$ **11.** $-4x^2$
13. $6x^9$ **15.** $-27k^5$ **17.** a^9 **19.** t^7 **21.** $r^{12}s^{12}$ **23.** $-a^6b^6c^4$
25. p^8q^8 **27.** u^{12} **29.** $3^6x^6y^{12}z^{18}$ **31.** $-u^{12}$ **33.** 81   **35.** x^3
37. m^2n^2 **39.** a^4b^2

B3.3, PAGES 195 – 196

1. a^5b^3 **3.** m^5n^7 **5.** $r^4s^5t^3$ **7.** $-u^8v^2$ **9.** $-72b^9c^8$

11. $5^47^3x^{15}y^{23}$ **13.** $st^8u^{23}v^{18}$ **15.** $2p^3$ **17.** $\dfrac{2q^2}{3}$ **19.** r^3s^2

21. c^4d^3 **23.** $\dfrac{2}{a^2}$ **25.** $\dfrac{1}{3b^5}$ **27.** $\dfrac{5x^3}{6y^2}$ **29.** $\dfrac{-4n^3}{m^3}$

31. $\dfrac{9u^4}{25v^6}$ **33.** $\dfrac{x^5}{y^7}$ **35.** $\dfrac{1}{p^7}$ **37.** $\dfrac{4a^2b^7c}{27}$

B3.4, PAGES 199 – 200

1. $\dfrac{1}{25}$ **3.** $\dfrac{1}{2}$ **5.** $\dfrac{8}{9}$ **7.** N.D.   **9.** 1   **11.** N.D.   **13.** 1

15. $\dfrac{1}{81}$ **17.** $\dfrac{x}{y^2}$ **19.** $\dfrac{k}{81}$ **21.** $\dfrac{3}{4}$ **23.** $\dfrac{1}{6}$ **25.** $\dfrac{1}{9x^3}$

27. $\dfrac{b^4}{16a^3}$ **29.** $\dfrac{1}{xy}$ **31.** m **33.** $\dfrac{v^4w^5}{u^2}$ **35.** $\dfrac{d^3}{c}$ **37.** $\dfrac{8b^5d^2}{9c^4}$

39. $29x^3y^6$

B3.5, PAGES 203 – 204

1. 8 **3.** $\dfrac{8}{27}$ **5.** $\dfrac{9}{16}$ **7.** $\dfrac{64}{25}$ **9.** $\dfrac{81}{64}$ **11.** 1 **13.** 64

15. $\dfrac{x}{y}$ **17.** $\dfrac{1}{p^6q^4}$ **19.** $\dfrac{x^6}{12^3}$ **21.** $\dfrac{m^9}{8n^6}$ **23.** $\dfrac{1}{4^5s^5}$ **25.** $\dfrac{a^4}{25b^6}$

27. $\dfrac{x^2z^4}{y^2}$ **29.** $\dfrac{r}{s^3}$ **31.** $\dfrac{1}{v^8w^{16}}$ **33.** 1 **35.** $\dfrac{1}{a^3}$ **37.** $\dfrac{81x^2}{64}$

39. $\dfrac{1}{ab^8}$

B3.6, PAGES 207 – 208

1. 5 **3.** $\dfrac{9}{4}$ **5.** $\dfrac{25}{3}$ **7.** $\dfrac{y}{9x^3}$ **9.** $\dfrac{a^2}{b^3}$ **11.** $\dfrac{q^3}{5p^2}$ **13.** $\dfrac{1}{pq}$

15. $\dfrac{1}{x^4y^3}$ **17.** $\dfrac{y^2}{x^2}$ **19.** 49 **21.** $\dfrac{a^2}{c^4}$ **23.** $\dfrac{q^2}{p^6r^2}$ **25.** $\dfrac{x}{3yz}$

27. $\dfrac{1}{5}$ **29.** $\dfrac{1}{x^3}+\dfrac{1}{y^3}$ **31.** $\dfrac{1}{(x-y)^2}$

B3.7, PAGES 211 – 212

1. 3.65×10^2 **3.** 5.28×10^3 **5.** 1.052×10^2 **7.** 2.12×10^{-3}
9. 6.25×10^{-2} **11.** 8.73×10^{-7} **13.** 1.414×10^0
15. 2.71828×10^{10} **17.** 3.75×10^{-1} **19.** 4.5×10^{-1}
21. 6.25×10^{-2} **23.** 9.175×10^1 **25.** 2.43×10^2
27. 2.401×10^3 **29.** 3.14159 **31.** 1.732 **33.** 650,000
35. 3.28×10^2 **37.** 6.4×10^{-3} **39.** 3.3×10^{-2}

ACHIEVEMENT TEST — B3, PAGES 213 – 216

1. D; $3 \cdot 3^2 \cdot a^3 \cdot a = 3(3^2a^{3+1}) = 3(3a^2a^4) = 3(3a^2)^2$
2. C; $a^2 \cdot a \cdot a \cdot b \cdot b^2 \cdot b \cdot c \cdot c^3 = a^{2+1+1} \cdot b^{1+2+1} \cdot c^{1+3} = a^4b^4c^4 = (abc)^4$
3. D; $3^{3+7} = 3^{10}$ **4.** B; $5^{3 \cdot 4} = 5^{12}$
5. E; $3^2(x^2)^2(y^3)^2 = 9x^{2\cdot2}y^{3\cdot2} = 9x^4y^6$

6. B; $\dfrac{6}{2}m^{6-2} = 3m^4$ **7.** A; $\dfrac{3}{12} \cdot \dfrac{a^{4-2}}{b^{4-3}} = \dfrac{1}{4} \cdot \dfrac{a^2}{b} = \dfrac{a^2}{4b}$

8. A; $v^3w^6v^4w^2 = v^{3+4}w^{6+2} = v^7w^8$

9. D; $\left(\dfrac{4xy}{3z}\right)\left(\dfrac{2z}{x}\right)^3\left(\dfrac{x^2}{8y^2}\right) = \dfrac{4xy \cdot 2^3z^3 \cdot x^2}{3z \cdot x^3 \cdot 8y^2} = \dfrac{4 \cdot 8 \cdot x^3yz^3}{3 \cdot 8 \cdot x^3y^2z} = \dfrac{4x^{3-3}z^{3-1}}{3y^{2-1}}$

10. E; $1 \cdot 6^2$ **11.** E; $\dfrac{3x}{y^2}$ **12.** C; $\left(\dfrac{4}{3}\right)^3 = \dfrac{4^3}{3^3}$

13. A; $\dfrac{b}{a^4} \cdot \dfrac{a^3}{b^5} = \dfrac{1}{a^{4-3}b^{5-1}}$

14. E; $\left(\dfrac{3}{n^4}\right)^{-3} = \left(\dfrac{n^4}{3}\right)^3 = \dfrac{(n^4)^3}{3^3}$;   or $3^{-3}(n^{-4})^{-3} = \dfrac{1}{27}n^{-4(-3)} = \dfrac{1}{27}n^{12}$

15. B; $3 \cdot 1 + 3 \cdot 5 = 3 + 15$

16. E; $\dfrac{2ab^2}{1} \cdot \dfrac{3b}{a^3} \cdot \dfrac{1}{4^2b^4} = \dfrac{6ab^3}{16a^3b^4} = \dfrac{3}{8a^{3-1}b^{4-3}}$

17. C; $\dfrac{4^2v^3v^4}{2^3u^2u^0} = \dfrac{16v^7}{8u^2}$ **18.** B   **19.** C; $0.0005 = 5.0 \times 10^{-4}$

20. C; $(96.9 \div 32.3) \times \left(\dfrac{10^{-2}}{10^5}\right) = 3.0 \times 10^{-2-5} = 3.0 \times 10^{-7}$

B4.1, B4.2, PAGES 219 – 220

1. 8   **3.** x^2 **5.** w^{16} **7.** $4x^7$ **9.** $10a^5$ **11.** $5x^6\sqrt{3}$

13. $2a^9\sqrt{5}$ **15.** $3a\sqrt{3}$ **17.** $10a^{15}\sqrt{3}$ **19.** $2p^{12}\sqrt{3}$ **21.** $2b$

23. $3x^3$ **25.** $5x^4$ **27.** $6a^3$ **29.** $6w^2x^2$ **31.** $\dfrac{3}{4}$ **33.** $6a^3$

35. $2x^2$ **37.** $5b$ **39.** $a + b$ **41.** $3a^2b\sqrt{3}$ **43.** $\dfrac{6ab}{7}$

B4.3, B4.4, PAGES 223 – 224

1. 6   **3.** 24   **5.** 8   **7.** $10x$ **9.** $x - 4\sqrt{x} + 4$ **11.** $b + 2b^2$

13. $a\sqrt{a}(a - 1)$ **15.** $5\sqrt{2}$ **17.** b **19.** 1   **21.** 7   **23.** 9

25. $6\sqrt{2}$ **27.** 14   **29.** $\sqrt{a}(2 + a)$ **31.** $b(\sqrt{b} + 2)$

33. $3\sqrt{x}(x + 2)$ **35.** $x(\sqrt{5} - \sqrt{3x})$ **37.** $\sqrt{x}(5 + x)$

39. $3x(\sqrt{x} - \sqrt{y})$

B4.5, PAGES 227 – 228

1. $a = 4$ **3.** $a = 2$ **5.** $x = 25$ **7.** $x = 4$ **9.** $x = 3$

11. $x = 10$ **13.** $x = 4$ **15.** $x = 8$ **17.** $x = 4$ **19.** $b = 4$

21. $x = 16$ **23.** $x = 1{,}296$ **25.** $k = 1$ **27.** $p = 36$ **29.** $x = 0$

B5.1, PAGES 231 – 232

1. $3x + 3 = 0$ **3.** $-12s + 3 = 0$ or $12s - 3 = 0$ **5.** $10n - 17 = 0$

7. $-5x - 3 = 0$ or $5x + 3 = 0$ **9.** $x = 3$ **11.** $s = -5$

13. $n = \dfrac{8}{3}$ **15.** $m = \dfrac{11}{7}$ **17.** $x = \dfrac{-15}{14}$ **19.** $s = \dfrac{22}{7}$

21. $k = 4$ **23.** $x = \dfrac{28}{17}$ **25.** $r = \dfrac{18}{25}$ **27.** 21 years old

29. 6

B5.2, PAGES 235–236

1. $x = \dfrac{1}{a}$ **3.** $t = \dfrac{1}{b}$ **5.** $y = \dfrac{4 + 5b}{3a}$ **7.** $s = -2$

9. $C = \dfrac{1}{2}F - 15$ or $C = \dfrac{1}{2}(F - 30)$ **11.** $h = \dfrac{s}{2\pi r} - r = \dfrac{s - 2\pi r^2}{2\pi r}$

13. $P = \dfrac{R}{1 + rt}$ **15.** $t = \dfrac{9}{15c - 8}$ **17.** $y = \dfrac{2}{9x - 3z} = \dfrac{2}{3(3x - z)}$

19. $r = \dfrac{b}{3b + 1}$ **21.** $b = \dfrac{-3ac}{a - 2c}$ **23.** $x = \dfrac{b}{-5a - 6} = \dfrac{-b}{5a + 6}$

25. $s = \dfrac{c + 2}{6c^2 - 2}$ **27.** A

B5.3, PAGES 239–240

1. $\dfrac{3}{7}$ **3.** $\dfrac{5}{5}$ **5.** $\dfrac{16}{14}$ **7.** yes, $\dfrac{54}{36} = \dfrac{57}{38}$ **9.** $x = 49$

11. 993.6 nautical miles **13.** 58 coupes **15.** 3600 quarts
17. 304 cups **19.** 101 dinners **21.** B = 6 **23.** 2625 revolutions
25. T = 3

B5.4, PAGES 245–246

1. $x = 1$ **3.** $s = \dfrac{1}{2}$ **5.** $x = -1$ **7.** $n = \dfrac{1}{2}$ **9.** $w = -2$

11. $x = -\dfrac{1}{4}$ **13.** $r = \dfrac{5}{33}$ **15.** $x = \dfrac{36}{11}$ **17.** $k = \dfrac{5}{22}$

19. $s = 3$ **21.** $t = -2$ **23.** $x = -\dfrac{11}{152}$ **25.** $z = \dfrac{-15}{23}$

B5.5, PAGES 249–250

1. $<$ **3.** $>$ **5.** $=$ **7.** $=$ **9.** $<$ **11.** $=$ **13.** $>$

15. $>$ **17.** $<$ **19.** $=$ **21.** $x < 3$ **23.** $x < 5$ **25.** $y \geq -2$

27. $r < -3$ **29.** $S \leq 2$ **31.** $t \leq 4$ **33.** $r > \dfrac{1}{5}$ **35.** $y \leq \dfrac{26}{17}$

37. $x \leq 30$ **39.** $n > \dfrac{2}{5}$ **41.** $k < \dfrac{69}{290}$ **43.** $r < 20$

B5.6, PAGES 255–256

1. $x = 1, y = 0$ **3.** $r = 2, s = 3$ **5.** $m = 1, n = 9$

7. $c = -2, d = 7$ **9.** $x = \dfrac{4}{5}, y = \dfrac{17}{5}$ **11.** $p = \dfrac{3}{2}, q = 5$

13. $x = \dfrac{29}{13}, y = \dfrac{-11}{13}$ **15.** $a = \dfrac{-48}{11}, b = \dfrac{-53}{11}$

17. $m = \dfrac{68}{39}, n = \dfrac{64}{39}$ **19.** Polly, \$27; Esther, \$23 **21.** 60 students

B5.7, PAGES 259 – 260

1. $x = 3$, $y = 2$ **3.** $r = -\dfrac{1}{2}$, $s = -\dfrac{13}{6}$ **5.** $m = 6$, $n = 0$

7. $u = -8$, $v = -10$ **9.** A = 6, B = 5   **11.** $s = \dfrac{29}{3}$, $t = -3$

13. $x = \dfrac{64}{59}$, $y = \dfrac{-28}{59}$ **15.** $m = \dfrac{164}{53}$, $n = \dfrac{46}{53}$

17. $a = \dfrac{23}{4}$, $b = \dfrac{3}{4}$ **19.** R $= \dfrac{19}{13}$, S $= \dfrac{-5}{13}$ **21.** $x = \dfrac{49}{19}$, $y = \dfrac{-1}{19}$

23. $m = \dfrac{50}{43}$, $n = -\dfrac{2}{43}$

ACHIEVEMENT TEST — B4, B5, PAGES 261 – 266

1. D; $7^2 = 49$ **2.** C; $(12x^2)^2 = 12^2 \cdot (x^2)^2 = 144x^4$

3. E; $(4a^8)^2 = 4^2 \cdot (a^8)^2 = 16a^{16}$ **4.** B; $\left(\sqrt{3^2 \cdot 3}\right)\left(\sqrt{r^{24}}\right) = 3\sqrt{3}\,r^{12}$

5. E; $\left(\sqrt{6^2}\right)\left(\sqrt{x^6}\right)\left(\sqrt{y^2}\right)\left(\sqrt{z^2 \cdot z}\right) = 6x^3 yz\sqrt{z}$

6. A; $\sqrt{\dfrac{a^{10}b^8}{4c^6}} = \dfrac{\sqrt{a^{10}}\,\sqrt{b^8}}{\sqrt{4}\,\sqrt{c^6}} = \dfrac{a^5 b^4}{2c^3}$ **7.** D; $\sqrt{(x-4)^2} = x - 4$

8. A; $\sqrt{169b^4} = 13b^2$

9. C; $2\sqrt{72} = 2\sqrt{9 \cdot 8} = 2\sqrt{3^2 \cdot 2^2 \cdot 2} = 2 \cdot 3 \cdot 2\sqrt{2} = 12\sqrt{2}$

10. B; $x(\sqrt{x} - x\sqrt{x}) = x\sqrt{x}(1 - x)$

11. C; $\sqrt{4^2 t^4 \cdot t} + \sqrt{8^2 t^2 \cdot t} = 4t^2\sqrt{t} + 8t\sqrt{t} = 4t\sqrt{t}(t + 2)$

12. A; $(2s^2)(5s^3) = 10s^5$ **13.** D

14. B; $(\sqrt{6n})(\sqrt{6}) + (\sqrt{6n})(\sqrt{n}) = \sqrt{6 \cdot 6 \cdot n} + \sqrt{6 \cdot n \cdot n}$
$$= \sqrt{6^2 \cdot n} + \sqrt{6 \cdot n^2} = 6\sqrt{n} + n\sqrt{6}$$

15. D; $(7 - 2\sqrt{a})(7 - 2\sqrt{a}) = 7 \cdot 7 - 7 \cdot 2\sqrt{a} - 7 \cdot 2\sqrt{a} + 2\sqrt{a} \cdot 2\sqrt{a}$
$$= 49 - 14\sqrt{a} - 14\sqrt{a} + 2 \cdot 2 \cdot \sqrt{a^2} = 49 - 28\sqrt{a} + 4a$$

16. A; $\sqrt{5x} = 6 + 4$; $\sqrt{5x} = 10$; $(\sqrt{5x})^2 = 10^2$; $5x = 100$; $x = 20$

17. D; $2 + 2\sqrt{t} = 24$; $2\sqrt{t} = 22$; $\sqrt{t} = 11$; $(\sqrt{t})^2 = 11^2$; $t = 121$

18. E; $7s - 5s = 5 + 3$; $2s = 8$; $s = 4$

19. E; $6r + 6 = 3r - 3$; $6r - 3r = -3 - 6$; $3r = -9$; $r = -3$

20. B; $J + 11 = 3(J - 13)$; $J + 11 = 3J - 39$; $-2J = -50$; $J = 25$

21. D; $4x + 3T = 5x - 5T$; $3T + 5T = 5x - 4x$; $8T = x$; $T = \dfrac{x}{8}$

22. C; $5an = 4 + 3a$, $\dfrac{5an}{5a} = \dfrac{4 + 3a}{5a}$, $n = \dfrac{4 + 3a}{5a}$

23. A; $\dfrac{3}{P} = \dfrac{5}{X}$, $3X = 5P$, $X = \dfrac{5P}{3}$

24. D; $X = \dfrac{12.5}{Y}$, $5 = \dfrac{12.5}{Y}$, $5Y = 12.5$, $Y = 2.5 = 2\dfrac{1}{2}$

25. B; $3x - 3x^2 = 1 - 3x^2$, $3x = 1$, $x = \dfrac{1}{3}$

26. E; $5x + 6 = 4(3x - 2)$, $14 = 7x$ **27.** C; $7r \le 28$

28. C; $2(3s + 2t) = 2(0)$ so $6s + 4t = 0$. $-3(2s - 3t) = -3(13)$ so $-6s + 9t = -39$. Add. $13t = -39$, $t = -3$.

29. D; From second equation, $y = 11 - 2x$. Substitute. $x + 2(11 - 2x) = 7$. $-3x = -15$, $x = 5$.

30. B; $S = 2J$, $S + J = 54$. Add $S - 2J = 0$ to $S + J = 54$, $3J = 54$, $J = 18$.

B6.1, PAGES 269 – 270

1. $x = 2$, $x = -3$ **3.** $a = 6$, $a = -3$ **5.** $b = 2$, $b = -2$

7. $x = 6$, $x = -6$ **9.** $m = 1$, $m = -1$ **11.** $w = -6$, $w = -8$

B6.2, PAGES 273 – 274

1. $x = -1$ **3.** $a = 2$ **5.** $x = 3, -1$ **7.** $w = 2, -4$ **9.** $x = 3$

11. $x = \dfrac{2}{5}$

B7.1, PAGES 277 – 278

1. E   **3.** F   **5.** H   **7.** A   **9.** A   **11.** B   **13.** D   **15.** A

17. N   **19.** P or Q   **21.** O   **23.** P   **25.** B   **27.** C   **29.** E

31. A

B7.2, PAGES 281 – 282

1.

3.

5.

7.

9.

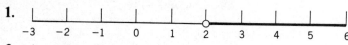

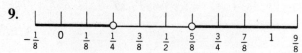

11. $x > 1$ and $x < 5$ **13.** $x > \dfrac{1}{8}$ and $x < \dfrac{1}{2}$ **15.** $x > -\dfrac{1}{8}$ and $x < \dfrac{1}{8}$

17. $x > -3$ and $x < -2$ **19.** $x > \dfrac{1}{2}$ and $x < 2$

B7.3, PAGES 285 – 286

1. $\left(2\frac{1}{2}, 3\right)$ **3.** $\left(-4\frac{1}{2}, 5\right)$ **5.** $(-3, -3)$ **7.** $\left(-2, \frac{1}{2}\right)$

9. $(5, -5)$. $\left(-6, -3\frac{1}{2}\right)$ **13.** $\left(2\frac{1}{2}, -3\right)$ **15.** $(-3, 4)$

17. -35.

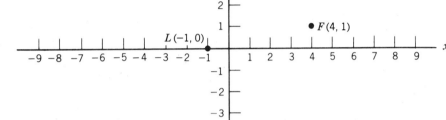

B7.4, PAGES 291 – 292

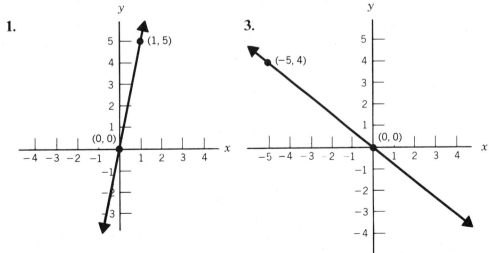

5. $y = 3x$

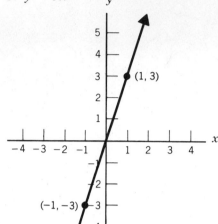

7. $y = 3x$

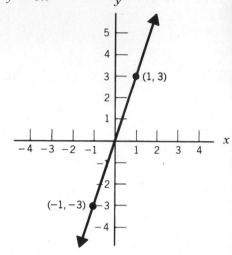

9. $x = 3$

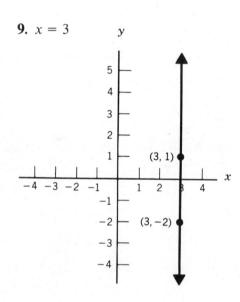

11. $y = \dfrac{3}{2}$

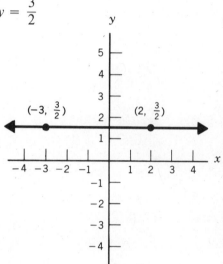

B7.5, PAGES 295 – 296

1. 2975   **3.** 3500   **5.** 48.1%   **7.** 65%   **9.** 6 hours   **11.** 19.2%
13. 21 women   **15.** 47 women   **17.** 34 women   **19.** 72.1%

ACHIEVEMENT TEST — B6, B7, PAGES 297 – 302

1. B; $5x - 10 = 0$, $x = 2$; $x + 3 = 0$, $x = -3$

2. B; $5x - 2 = 0$, $x = \dfrac{2}{5}$; $4x + 3 = 0$, $x = -\dfrac{3}{4}$

3. A; $(2x)(3x) - 1(3x) - 2(2x) - 1(-2) = 0$, $6x^2 - 7x + 2 = 0$

4. A; $(2x - 3)(x + 1) = 0$; $x + 1 = 0$, $x = -1$; $2x - 3 = 0$, $x = \dfrac{3}{2}$

5. D; $(2s - 3)(s + 1) = 0$; $2s - 3 = 0$, $s = \dfrac{3}{2}$; $s + 1 = 0$, $s = -1$

    6. B   **7.** E   **8.** A   **9.** C   **10.** C   **11.** D   **12.** B
    13. E   **14.** B   **15.** A; 25% + 12% = 37%   **16.** A; 33 + 10 = 43

C1.1, PAGES 309 – 310

Problems 1–5: Answers will vary
7. a. 400 in.   b. 16.45 in.   c. 6.70 in.   d. 17.17 in.   e. 56.42 in.
    f. 49.49 in.

C1.2, PAGES 313 – 314

1. 11.4   **3.** 37.8   **5.** 29   **7.** 10.5   **9.** 31.2   **11.** $P = 2l + 2w$, 20
13. $P = a + b + c$, 21.6   **15.** $P = 8s$, 70.4   **17.** $2l + 2w$, 19.4
19. $P = 4s$, 92.8

C1.3, PAGES 319 – 320

1. 6.25 square units   **3.** 12 square units   **5.** 20 square units
7. 715 square units   **9.** 4.2 square units   **11.** 32 square units
13. 6   **15.** 5

C2.1, PAGES 323 – 324

1. 50.24   **3.** 28   **5.** 35.325   **7.** 37.68   **9.** 2.87   **11.** 12
13. 28.1

C2.2, PAGES 327 – 328

1. D = 16, C = 50.24, A = 200.96   **3.** R = 19, C = 119.32, A = 1133.54
5. R = 3.99, D = 7.98, C = 25.06   **7.** R = 2, C = 12.56, A = 12.56
9. R = 0.56, D = 1.12, C = 3.52   **11.** 144π or 452.16 square units
13. 169 square units   **15.** 1 to 4, or $\dfrac{1}{4}$

C3.1, PAGES 331 – 332

1. 240 cubic units   **3.** 64 cubic units   **5.** $\dfrac{3}{64}$ cubic units
7. $\dfrac{1}{125}$ cubic units   **9.** 1728 cubic units   **11.** 36 cubic units
13. yes   **15.** 0.125 cubic unit

C3.2, PAGES 335 – 336

1. 225π ft^3 **3.** 972π in^3 **5.** 27π in^3
7. The cylinder volume is $\dfrac{3}{4}$ of the sphere volume.
9. The volume doubles   **11.** 27 times

ACHIEVEMENT TEST — C1, C2, C3, PAGES 337–340

1. B; $2l = P - 2w = 12$, $l = 6$. $A = l \times w = 12$

2. B; $l = A \div w = 8$; $P = 2l + 2w = 16 + 8 = 24$

3. A; $w = A \div l = 16 \div 16 = 4$

4. A; $PQ \times TS = A$. Thus, $TS = A \div PQ = 36 \div 9 = 4$

5. E; Area of triangle BCE = Area of triangle ABE = $\frac{1}{2} \cdot 3 \cdot 3 = 4\frac{1}{2}$

Area of triangle CDE = Area of triangle ADE = $\frac{1}{2} \cdot 5 \cdot 3 = 7\frac{1}{2}$

The area of ABCD is $4\frac{1}{2} + 4\frac{1}{2} + 7\frac{1}{2} + 7\frac{1}{2} = 24$

6. D; $\frac{1}{2}bh = A$, $\frac{1}{2} \cdot 9 \cdot XY = 45$; $XY = \frac{2}{9} \cdot 45 = 10$

7. C; Doubled length is 12. Doubled width is 4. $12 \times 4 = 48$.

8. A; $\frac{1}{2}bh = A$, $\frac{1}{2} \cdot 20 \cdot XC = 60$, $XC = 60 \div 10 = 6$

9. E; Diameter = Side = $16 \div 4 = 4$. Radius = Diameter $\div 2 = 2$.

10. B; $C = 2\pi r = 16\pi$. $R = 16\pi \div 2\pi = 8$. $A = \pi r^2 = \pi(8^2) = 64\pi$.

11. C; $A_1 = 6^2\pi = 36\pi$. $A_2 = 12^2\pi = 144\pi = 4 \times 36\pi = 4 \times A_1$.

12. E; $P = DC + AB + \frac{1}{2}(2\pi r) + \frac{1}{2}(2\pi r) = 6 + 6 + 3\pi + 3\pi = 12 + 6\pi$.

13. E; $6 \times 6 \times 6 = 216$ **14.** B; $3 \times 5 \times 8 = 120$

15. C; $\pi r^2 h = \pi \cdot 4^2 \cdot 9 = 144\pi$ ft^3 **16.** A; $V = \frac{4}{3}\pi r^3 = \frac{4}{3} \cdot 3^3 \cdot \pi = 36\pi$

17. D; $V = 10 \times 6 \times 8 = 480$

18. C; $V_2 = \frac{4}{3}\pi(2r_1)^3 = \frac{4}{3}\pi \cdot 8r^3 = 8\left(\frac{4}{3}\pi r^3\right) = 8V_1$

C4.1, PAGES 343–344

1. Answers may vary   **3.** All three sides should have the same length.

5. yes, yes   **7.** no, yes   **9.** The angles are the same.

C4.2, PAGES 347–348

1. 39°   **3.** 100°

5. $\angle A = 26\frac{2}{3}^\circ = 26°40'$, $\angle B = 56\frac{2}{3}^\circ = 56°40'$, $\angle C = 96\frac{2}{3}^\circ = 96°40'$

7. 45°   **9.** no   **11.** 360°

C5.1, PAGES 351–352

1. $\angle B = 50°$, $\angle C = 80°$ **3.** $\angle Q = 25°$, $\angle R = 130°$

5. $\angle F = 146°$, $\angle G = 17°$ **7.** $\angle M = 80°$, $\angle N = 50°$

9. $\angle X = 35°$, $\angle Y = 110°$

C5.2, PAGES 355 – 356

1. 10.8   **3.** 15   **5.** 2 or 4, 360   **7.** 6 or 4, 720   **9.** 8, 1080

C6.1, PAGES 359 – 360

1. yes, SSS   **3.** yes, SAS   **5.** yes, AAS

C6.2, PAGES 363 – 364

1. $\angle A \cong \angle X$, $\angle B \cong \angle Y$, $\angle C \cong \angle Z$ **3.** 16   **5.** $9\frac{3}{5}$ or 9.6

7. $1\frac{1}{2}$ or 1.5   **9.** 3 to 4, or $\frac{3}{4}$

ACHIEVEMENT TEST — C4, C5, C6, PAGES 365 – 368

1. D

2. A; $\angle X = \angle Z$ and $\angle X + \angle Z + 30° = 180°$. $2 \times \angle X = 150°$. $\angle X = 75°$.

3. B; $180 - (30 + 30) = 180 - 60 = 120$

4. D; $\angle DCB = 90°$. Thus, $\angle BDC = 180 - (50 + 90) = 180 - 140 = 40$

5. D   **6.** B

7. A; $\angle A = 60°$, $\angle B = 60°$, $\angle C = 60° + 60° = 120°$

$\angle D = 60° + 60° = 120°$.

$\angle A + \angle B + \angle C + \angle D = 60° + 60° + 120° + 120° = 360°$

8. A; $\angle XZY = 60°$. Thus, $\angle YZW = 180° - 60° = 120°$. YZ = ZW because YZ = XZ and XZ = ZW. Thus, triangle YZW is isosceles. Therefore, $\angle ZYW = \angle W$. $\angle W = \frac{1}{2}(180 - 120)° = \frac{1}{2}(60)° = 30°$.

9. B; $\angle PRQ = 180° - 110° = 70°$. $\angle P = \angle PRQ = 70°$. Thus, $\angle PQR = 180 - (70 + 70) = 40°$.

10. E; One side of the hexagon = one side of one triangle = $\frac{1}{3}\sqrt{5}$.

$P = 6\left(\frac{1}{3}\sqrt{5}\right) = 2\sqrt{5}$.

11. B; $\angle BCD = 180° - (50° + 20°) = 110°$. $\angle DCA = 180° - 110° = 70°$.

12. D; $\dfrac{\overline{VY}}{VW} = \dfrac{XY}{XZ}$, $\dfrac{3}{5} = \dfrac{XY}{10}$, $5XY = 30$, $XY = 6$.

13. D; If AB = XY, ASA is satisfied.

14. E; $\dfrac{3}{DE} = \dfrac{5}{6}$, $5DE = 18$, $DE = 3.6$

15. B; $\dfrac{6}{10} = \dfrac{T}{60}$, $10T = 360$, $T = 36$

C7.1, PAGES 373 – 374

1. 289, 64, 225   **3.** 169, 25, 144   **5.** 625, 400, 225   **7.** 15
9. $\sqrt{6}$ or 2.4   **11.** 5   **13.** $\sqrt{41}$ or 6.4   **15.** $\sqrt{26}$ or 5.1   **17.** 6

C7.2, PAGES 379 – 380

1. $9\sqrt{2}$ **3.** 5   **5.** $15\sqrt{2} \approx 21.21$ **7.** $\sqrt{2}$ to 4, or $\dfrac{\sqrt{2}}{4}$ **9.** 8
11. $8\sqrt{3} + 8$

C8.1, PAGES 383 – 384

1. a and b, x and y **3.** a and b; c and d; e and f; g and h
5. 128°: $\angle a$, $\angle b$, $\angle e$, $\angle f$.   52°: $\angle c$, $\angle d$, $\angle g$, $\angle h$ **7.** 95°   **9.** 131°

C8.2, PAGES 387 – 388

1. a. $\dfrac{2}{3}$ b. $-\dfrac{3}{2}$ **3.** $l_1 \perp l_3$ **5.** no   **7.** 30°   **9.** $\sqrt{55}$, $\sqrt{73}$
11. $-\dfrac{8}{3}$ **13.** 37°, 53°, 90°   **15.** $5^2 + 8^2 \neq 9^2$

ACHIEVEMENT TEST — C7, C8, PAGES 389 – 390

1. B; $c^2 = 5^2 + 2^2 = 29$, $c = \sqrt{29}$
2. B; $(2\sqrt{5})^2 + a^2 = (\sqrt{30})^2$, $a^2 = 30 - 20 = 10$, $a = \sqrt{10}$
3. B; $c^2 = 7^2 + (\sqrt{15})^2 = 49 + 15 = 64$, $c = \sqrt{64} = 8$
4. B; $(XZ)^2 + 11^2 = (\sqrt{219})^2$, $(XZ)^2 = 219 - 121 = 98$, $XZ = \sqrt{98} = 7\sqrt{2}$.
5. D; $PR = 2(QR) = 2(2\sqrt{2}) = 4\sqrt{2}$
6. A; At the 90° angle, label the point S. $1^2 + (PS)^2 = (\sqrt{17})^2$, $(PS)^2 = 16$,
    $PS = 4$. $1^2 + (SR)^2 = (\sqrt{10})^2$, $(SR)^2 = 9$, $SR = 3$. $PR = PS + SR = 4 + 3 = 7$.
7. D; $C^2 = 6^2 + (3\sqrt{5})^2 = 36 + 45 = 81$, $c = 9$.
8. C; $180° - 96° = 84°$
9. A; The triangle is a right triangle, so the top angle is 60°. $x = 180° - 60° = 120°$.

INDEX